Praise for Stuart Clark's trilogy

The Sky's Dark Labyrinth

'A vivid, thrilling portrayal of the lives and work of Kepler and Galileo . . . Books like this transform the way we access and understand our view of history'
Lovereading UK

'Sit under the stars and wonder, not just at their eternal beauty and mystery, but at the courage of the men who risked their lives so we could understand them'
Daily Mail

'Usually when reading a novel based on historical figures and true events, I find myself at some point asking what it's based on. It's a testament to Clark's ability to tap into the seventeenth-century mindset that this time the question never arose'
Historical Novel Society

'I could all but smell the streets and markets of seventeenth-century Prague in this novel'
Nature

The Sensorium of God

'Stuart Clark has once again invoked the great debates using enough science to help the reader's understanding without over-blinding laymen like me. The concluding volume will be about Einstein. My breath is bated'
Kathy Stevenson, *Daily Mail*

'Clark does a sterling job of covering the tricky period when scientists were the superstars of society'
New Scientist

'The best historical fiction goes beyond dates and events, giving historical figures emotions, achievements and failings. This is very much the case here, where petty squabbles sit beside philosophical debate in a rounded picture of great men and ideas'
We Love This Book

STUART CLARK

Stuart Clark's career is devoted to presenting the complex world of astronomy to the public. He holds a first-class honours degree and a PhD in astrophysics. He is a Fellow of the Royal Astronomical Society and a former Vice Chair of the Association of British Science Writers. In 2000 *The Independent* placed him alongside Stephen Hawking and the Astronomer Royal, Professor Sir Martin Rees, as one of the 'stars' of British astrophysics teaching.

He divides his time between writing books and his blog, Across the Universe, for the Guardian. He is a consultant to *New Scientist*, writes features for BBC *Focus* and BBC *Sky at Night* and is a former editor of Astronomy Now magazine. Until 2012, he was senior editor for space science at the European Space Agency.

www.stuartclark.com

The Sensorium of God

Stuart Clark

THE SKY'S DARK LABYRINTH TRILOGY
Book II

Polygon

First published in hardback in Great Britain in 2012
by Polygon, an imprint of Birlinn Ltd. This paperback edition published in 2013 by Polygon.

Birlinn Ltd
West Newington House
10 Newington Road
Edinburgh
EH9 1QS

www.polygonbooks.co.uk

ISBN 978 1 84697 237 9
eBook ISBN 978 0 85790 079 1

British Library Cataloguing-in-Publication Data
A catalogue record for this book is available on request from the British Library.

Typeset by IDSUK (DataConnection) Ltd
Printed and bound by Clays Ltd, St Ives plc, Bungay

Content approved by NMSI Enterprises/Science Museum.
Licence no. 0283.

PART I
Action

1
Woolsthorpe, England
1679

There was no place for daylight in this morbid womb. The alchemist smoothed the heavy curtains across the window and pressed them into the corners of the frame, determined to banish any sunshine from the room. Candles would provide all the illumination he needed to perform his work.

Edging round a table cluttered with bottles and vials, he approached the bed. His eyes came to rest on its unconscious occupant, whose outline was scarcely visible beneath the heap of blankets.

He fumbled in his pocket and pulled out a small mirror. Just a fortnight ago he had used it to bounce light across his rooms in college. Now, he held it close to his mother's mouth. Reassured by the faint condensation that collected on the polished metal, he straightened. His eyes stung from exhaustion.

Returning to the window, he stood in front of the table and stared at the bottles. Each one contained something different: a salt, a herb, some liquid essence. Now they were all that stood between his mother and death. He drew his hands together as if in prayer and raised them to his face. Taking a sharp breath, he shoved the baggy cotton shirtsleeves up his thin arms and set to work.

He picked up a flask, upturned it and shook the final drops of a previous concoction to the floor. Then he began swirling a new mixture together.

Three drops of yellow dock root oil – to help purify the blood.

A sprig of saturated thyme – to bring courage.

Coffee essence – to stimulate the heart.

A pinch of crushed turmeric.

A few days ago he would have measured everything and kept track of the recipes in his notebook, but yesterday his mother had slipped into an unbreakable sleep. Now, he sloshed together anything that came to mind. He set the bottles chinking as he snatched up one container after another.

God grant her reprieve and I will atone for every wicked thought I have ever harboured against her, he vowed.

He took the flask and leaned over his mother to drip the curative on to her thin lips. As he did so, the door opened. Unwelcome light entered the room, making him blink.

'Come along, Isaac.'

Newton shot a piercing look at the all-too-familiar figure bustling into the room. The portly woman carried a tray of food.

'I must work,' he snapped.

'Nonsense.' Mrs Harrington thrust the tray at him.

Upon his arrival last week, he had found the troublesome woman supervising the household with misplaced industry.

'You must eat,' she said. 'There's no point in making yourself sick as well.' There was no warmth in her voice.

'What concern is it of yours? You don't belong here.'

'I'm your mother's closest friend. I've known you since you were born. I helped deliver you, in that very bed.'

'You were just a girl from the village, hired to help out during my mother's confinement. Only later did you worm your way into her affections.' Newton crossed to the table and lifted more bottles, inspecting their contents. 'Isn't it true that you were so convinced I wouldn't last my first night that you dallied in conversation when you were sent for medicines?'

She drew herself up. 'Your mother grows weaker every day, Isaac. It's time to say goodbye to her.'

'No! I haven't finished my ex–' He cut himself off.

'Your what? Your experiments?'

'My treatments, my treatments.'

Mrs Harrington softened her voice. 'Isaac, let her pass in peace, with sunlight and warmth.' She pushed past him, making for the curtains.

'Don't open the window. There are miasmas in the air, my mother is too frail.'

'She needs fresh air. It reeks in here.'

'The potions react with sunlight. You'll destroy them.'

She gave him a sour look. Balancing the tray on one hip, she pulled the curtain open. The sunlight cut into Newton's eyes. He gripped the sides of the table to steady himself against the molten force rising inside him.

She crossed to the mantelpiece, still carrying the tray. 'You think your mother wants all this?' She cast her disdainful gaze over more bottles and flasks.

'My mother will be treated with whatever I choose,' he said through clenched teeth.

A knowing smile crossed her face. 'I've watched you and your wicked little ways since you were a boy.'

'The boy you remember has grown up.'

'As far as I can see, nothing's changed. I remember you peering round the congregation while the rest of us bowed our heads in prayer. You always did think you were something special.'

Newton's hands jerked free from the table and he started groping among the bottles. He picked something up and without thinking hurled it in Mrs Harrington's direction.

Jaw dropping in disbelief, she heard it smash against the wall. Newton's second missile found its mark, glancing off her fleshy left cheek. She reeled backwards, dropping the contents of the tray across the floorboards, and fled the room, screaming.

Newton rushed after her, kicking the food aside and slamming the door shut with his whole body. He stumbled to the curtains and yanked them shut before returning to his mother's bedside. Her jaw was slack. He reached out to touch her face, hesitating at the last moment. When his ink-stained fingers made contact with the skin, terror flashed inside him and he snatched his hand away. She was gone. Tears of rage spilled from his eyes. When the spasm subsided, he sank to his knees and raised his face to heaven.

'Lord, why do you test me with such impossible tasks?'

The servants took Hannah's body and sewed it into a woollen burial garment and bonnet. Quiet tears accompanied their busy hands. They laid the body on a table in the parlour where Newton stood in one corner, keeping vigil.

Outside the world had turned to green, yet he seemed immune to its warmth. His clothing hung as if tailored for a larger man. To hide this fact he had swaddled himself in his academic gown, embracing its comfortable feeling of far-off Cambridge.

He could not go near the corpse. How small she looked compared to his memory. The bonnet covering her hair could not disguise the tangled white eyebrows and the thin skin stretched over her cheekbones. She was as pale as the coarse wool of her shroud, and with each passing hour looked less as though she had ever been alive.

He conjured the memory of her once delicate face, with its attentive eyes and the tumbling ringlets of dark hair that she had let him stroke. He tried to hold the image but another kept forcing it away. He saw his mother in the distance, peering out of the carriage that had taken her away from him for eight years. It was his earliest memory. The scene replayed itself in staccato images, and he watched for the millionth time as she broke her gaze and looked away from him long before she had to, before the vehicle had drawn out of sight.

Why had she turned away? The question haunted him, now perhaps more than ever before. *What if she were magically to open her eyes now?* The impossible reality filled his mind. She would sit upright, the sunlight from the window catching the outer threads of her woollen garments but rendering the rest of her in silhouette. She would talk in the softest of voices, her features indistinct, answering his questions without hesitation.

– Why did you leave me, mother?

– I fell in love with Barnabas. I wanted to start again, to pretend you never existed.

– Why?

– You always were difficult. You didn't need me or want me.

– Then why did you return when he died?

– We couldn't stay in the rectory once Barnabas had died. I needed a home for my new children. You just got in the way.

– Did you love them more than me?

– Of course I did.

Movement through the panelled window snapped him from his imaginings. The round ball of Woolsthorpe's rector was making its way up the path towards the front door.

Newton left the parlour and opened the front door before the Reverend Hazel could knock.

'My condolences, Mr Newton. Your mother was a credit to the parish.' Hazel made to step forward, stopping himself just in time to avoid walking straight into Newton's immobile form.

'You have a difficult task, Reverend. My mother has not left a will. I searched her papers today but could find nothing.'

Hazel patted his leather satchel. 'I have it here. Mrs Harrington delivered it to me at the beginning of the week.'

Suspicion flared in Newton. 'Let me see it.'

Hazel fumbled around, eventually producing a folded piece of paper. Newton seized it and unfolded the sheet. It was indeed covered in his mother's handwriting.

'As you see, you are sole executor and principal beneficiary,' said Hazel. 'She has left the house and estate to you in its entirety.'

Newton's eyes worked through the almost illegible scribble. Indeed, Woolsthorpe Manor was now his. The other gifts were trifling, some five pounds for the parish poor and similar sums for the serving staff.

'There is little provision for your siblings,' said Hazel tentatively.

'Half-siblings,' Newton corrected him. 'They inherited from their father. They want for nothing.'

Hazel paused before answering. 'If that is your decision.'

'It is.' Newton handed him back the will.

'Then there is just one other matter. Your mother indicates that she is to be buried according to your wishes.'

'In the churchyard, next to my father.'

Again, Hazel paused. 'As you wish. Well, I see no further reason to delay you.'

'I'm aware that you had a high regard for Barnabas Smith, Reverend, but I know my mother.'

'God bless you, Mr Newton.' The coldness in Hazel's voice made it sound like a curse.

Newton shut the door on his receding form and looked around the hallway, seeing the plaster and timbers anew. For thirty-seven years – his entire lifetime – he had either lived in or visited this house and walked its fields. Now it was his, yet he felt more at home in a small room overlooking the quadrangle in Trinity College. He grimaced at the irony.

The mourners gathered by the church. Dressed head to foot in black, they were silhouettes against the sun that had chosen to mock them with its rays.

Newton looked around and thought what a pitiful end it was. The servants and the farm hands were there, snuffling and grizzling. There were a few villagers and, of course, his half-brother and half-sisters. They stood together, Mary, the eldest, blinking tears and being comforted by Hannah, the youngest and his mother's namesake. Benjamin looked angry, as usual. He bristled as Newton greeted them.

'Not here, Ben,' whispered Hannah.

Newton scrutinised his half-brother and thought what an imperfect rendering he was: short and lacking a well-proportioned oval face, with a nose that was too small. The glare from his red-rimmed eyes, however forceful, inspired no confidence.

'Why bury her here?' Benjamin asked.

Newton should have guessed this would be his lament. 'To be with her true husband.'

'But you never knew him! He died just eighteen months after marrying her, while she was still carrying you. She was married to my father for eight years. If there is a rightful place–'

'Silence!' Newton raised his hand. 'I will hear nothing from the man who killed her.'

Benjamin's mouth gaped in astonishment.

'Isaac, how can you say that?' gasped Hannah.

'By nursing him – a grown man – my mother contracted the fever herself. It's nothing but a statement of fact.' Newton turned from their incredulous faces, exhilarated by the shock he left in his wake.

The congregation stood around the grave, as unmoving as the headstones, while Hazel mumbled his way through the eulogy. He insisted on referring to the body as Hannah Smith, a name so alien to Newton that his mind soon drifted, leaving the minister to work through his platitudes.

Left at Woolsthorpe in the care of his grandparents, Newton had never known when his mother would visit.

There was not enough consistency to her appearances to allow him to discern their shape; they could fall on any day of the week and last anything from a whole day to a meagre hour. He would present her with the latest of his wooden models, maybe a simple sundial or a mobile windmill, and she would politely contemplate the creation and then ask how his schooling was progressing.

When it was time for her to leave – always too soon – his grandparents would force him to watch her go. Sometimes the pain of her departure would rumble inside him like summer thunder; other times it would strike like a bolt of lightning. If he tried to run away or he succeeded in hiding, his grandmother would threaten at the top of her wavering voice, 'If you cannot behave yourself, we'll have to stop your mother coming to see you.'

'I wish she were dead,' Newton had screamed at them when he had heard the ultimatum once too often. In the shocked aftermath he had fled the house, returning at twilight, tear-stained and grubby but unrepentant.

For eight years this hateful cycle repeated itself, but then came the biggest tragedy of all. His mother returned for good, but trailing a meddlesome six-year-old girl, a weakling boy and a screaming baby. Daily, Newton was forced to witness her devotion to them while she all but ignored him.

Newton's thoughts returned to the present. He watched with utter disbelief as the small body was lowered into the grave, wincing when he noticed that Benjamin was helping the undertakers.

Where are you, Mother? With Father in heaven? With that devil Smith, suffering for abandoning me? Well, God cannot help you there. Only I can forgive you that sin.

The day after the funeral Newton called the servants to the kitchen. They stood in silence, straightening uniforms and tucking away curls of hair. Mrs Harrington rose from her throne at the end of the long, heavy table, her hand fluttering self-consciously to the bruise on her cheek.

'Mrs Harrington, your services are no longer required. You will leave immediately,' Newton barked.

She looked around for support but no one would meet her gaze. Swallowing hard, she regained her composure. 'You wicked man! What would your mother say?'

'My mother is no concern of yours. She has her sins to atone for, you have yours.' He lifted his dimpled chin to look down his long nose at her.

'She loved you, Isaac, more than you deserved. That's why she left you here when she married Smith. He wanted Woolsthorpe, but she insisted it was yours alone. He forbade her to bring you to live with him, thinking she'd relent. But she didn't, because she knew you'd always be overlooked in favour of any children she had with Smith. So she bound you up in Woolsthorpe so tightly that no one could contest it when she left it all to you.'

'We were happy here, together,' Newton heard himself say, and instantly chided himself for revealing his thoughts.

'You were three when she left. How could you possibly understand that your mother was struggling for money? She had to marry Smith. He was already old when she married him. It was just your bad luck that he lived until he was seventy-one.'

Newton's eyes locked with hers briefly, then turned to the gleaming pots and their curved reflections of the scene. 'You're trespassing. Leave my grounds and never return.'

When he glanced back, a half-smile played across Mrs Harrington's face. As she stalked to the kitchen door, she turned and said: 'One day, you will meet someone who will do to you what you have done to me. Mark my words.'

2
Southwark

A stooped figure paced the cobbles outside the Crown Inn, hands clamped behind its back, downcast face invisible beneath a shock of greasy hair. Only a sharp nose protruded, from which dripped the occasional dewdrop of clear fluid.

Most people gave the shuffling homunculus a wide berth even when the courtyard became crowded with carriages and new arrivals. They edged around him as if his crooked spine were contagious. It was a reaction that Robert Hooke was used to and, for the most part, ignored. After all, avoiding eye contact was the easiest thing for a hunchback to do. So he contented himself with watching his own feet, craning his neck only when the whirling clatter of approaching wheels and hooves drew his attention.

What would she look like?

Certainly not the vivacious young woman he had sent back to the Isle of Wight eighteen months earlier, if her mother – his sister-in-law – were to be believed. No, Grace would be broken and contrite. Perhaps he would put a protective arm around his niece as they walked home, to let her know that one member of the family still loved her.

The evening air invaded his ill-fitting jacket through the gaping cuffs and the threadbare patches at his elbows and shoulders. His stockings and voluminous breeches offered little protection from the chill either. He pulled the jacket tighter and paced around some more.

The colour was draining from the rectangle of sky above the courtyard. If the Portsmouth coach did not arrive soon,

Hooke would have to leave before the bridge gates were closed for the night and return in the morning.

Perhaps, he thought, *I could book a room and wait here.*

He checked the pouch containing his money and was about to head inside when a battered old coach rattled into the yard. Almost at once the passengers disembarked, stretching and helping each other down. They laughed and exchanged pleasantries as their luggage was unloaded. A young woman turned to leave the group.

It was not her.

This woman wore a velvet cape and walked with purpose, carrying a large bag with ease. He looked past her, hoping for another female to sidle out. Perhaps he should go and look inside; maybe she was afraid to face him. She had good reason.

'Hello, Uncle,' said the woman.

Hooke fought to straighten himself to look at the stranger.

'It's me,' she said.

'Grace?'

She smiled, a flash of white teeth behind cherry lips, and the breath left his lungs. He stared into her dark eyes. Where was the ruined girl he had been expecting? Grace was supposed to be cowed, not yet nineteen but her bloom already plucked. He had imagined her on the verge of tears, unable to look at him for shame.

She produced a handkerchief and dabbed his chin. A red spot appeared on the material, no doubt the result of his last-minute decision to shave before hurrying to Southwark. The gesture broke his paralysis. He swatted her away and said angrily, 'I can look after myself.'

She looked around in case others had witnessed the rebuff.

'You're not even sorry, are you?' he said.

'Uncle. . .'

He could hardly bring himself to look at her. 'From now on, you will address me as Mr Hooke. You are my housekeeper. Nothing more.'

They walked in silence to the bridge, where a large crowd was milling around the gatehouse. The imposing wooden structure towered above them, and people looked blankly from one to another. As was usual when confronted with a crowd, Hooke tried to burrow through, but this time the way was blocked by the press of people. The musty smell of unwashed clothing hung in the air.

A plump woman turned to stare down at him. 'You'll have to wait your turn like the rest of us.'

Hooke retreated. He could see nothing but people's backs. 'What's going on?'

Grace stood on tiptoe and peered into the crowded passage. The bridge was like a tunnel with wooden buildings lining either side. She saw people hanging out of top-floor windows, calling to each other and pointing at something, while at street level the shopkeepers hurried to gather their wares before they were trampled or stolen in the crush. 'Everything's at a standstill. I cannot see why.'

Hooke singled out a young man who had just emerged from the throng. 'You there, what's the hold-up?'

The man jerked a thumb over his shoulder. 'A cart's got a wheel off. You'll be lucky to get across before the hour's up.' He paused long enough to look Grace up and down, then hurried on his way.

'We'll take a wherry. Come on, before everyone has the idea,' said Hooke irritably.

They cut through the crush of people into one of the alleys that ran parallel to the riverbank. It was dark now, and Hooke stepped up his pace. As they headed downstream, away from the torrents of water cascading between the wooden bulwarks of the bridge, Hooke's breathing became laboured.

'We can slow down if you want,' said Grace.

'I will walk at whatever speed I choose,' he said, but slowed down nevertheless.

It grew cooler as they descended a flight of stone steps to the riverside. Shadows moved on the quayside and the lanterns on a line of river taxis bobbed up and down.

'Need a ride?' asked a gruff voice.

'North bank,' said Hooke.

The wherryman caught sight of Grace and quickly reached for her bag. 'Allow me, miss.' His voice had softened.

'Thank you,' she purred.

Hooke's blood almost boiled.

The man stowed Grace's bag, then he took her gloved hand and guided her into the boat. For a moment, they looked as if they were dancing.

Hooke plunged from the bank, setting the small vessel rocking. Unbalanced by the motion, he sat down heavily.

'Are you all right, Uncle?'

Hooke glared.

'Sorry . . . Mr Hooke.'

'Perfectly,' he spat. 'Now, let's be on our way.'

'Oh yes, sir,' the wherryman said with just a hint of sarcasm. He cast off and pumped the oars with his big, muscled arms, straining against the river. As he rowed, he stole frequent glances at Grace.

The towering shape of the bridge loomed to one side, almost lost to view except for the occasional lighted window. They cut through the inky water, navigating the constellation of boat lights that danced around them as others crossed the river. When they reached the far bank the wherryman set the oars, ensuring that they landed with nothing more than a soft tap on the quayside.

Dismissing the man with some coins, Hooke turned to Grace. 'Hurry up, girl,' he grumbled.

They headed northwards, leaving behind the crowds and entering the hinterland between the stone of rebuilt London and the older realms of wood and plaster. Half-finished buildings lined their way, deserted now that the workmen had left for the day. In the darkness, it was easy to confuse the nascent buildings with ruins.

Hooke crossed the road, pacing out its width from force of habit. Anger flashed as he reached the opposite kerb; the road was at least two feet narrower than his written specification.

Why did the authorities insist on squeezing London back into its mediaeval claustrophobia? Did no one ever learn? The labyrinthine old alleys and shanties had proven to be coffins when the fire came. Nursing his annoyance, he even forgot to watch the corners for pickpockets.

After what seemed like an age, with Grace trailing along silently behind, the familiar bulk of Gresham College loomed. A dozen years ago, it had been on the safe side of the Duke of York's firebreaks and escaped the flames. It had become a sanctuary for city officers and financiers. Now, however, it found itself painfully unfashionable. The flesh was gone from its timbers, leaving just the matted sinew of the wood to hold the structure upright. Yet, for all its obvious antiquity, it was home.

Hooke fumbled with his keys and the side door creaked open. He led Grace through a shabby hallway and back outside into the quadrangle. They crossed to the far corner.

Hooke's rooms were little warmer than outside. He wasted no time getting a collection of apple logs burning in the grate, then set about kindling the rushlights dotted around the main room. As the fire drove away some of the cold and the strong odour of rat droppings, Grace circled the large room. Her footsteps set the floorboards creaking beneath the threadbare rugs. She stepped over the pieces of abandoned apparatus and ran her fingers along the wood panelling. She straightened a few of the portraits and lingered by the fire. 'It's good to be back.'

Hooke buried the shred of pleasure provoked by her comment.

'Am I to sleep in the turret room?' she asked.

He gave a curt nod.

With a rushlight in one hand and her bag in the other, she disappeared through the door that led to the turret.

'Don't be long,' Hooke called. 'There are provisions in the kitchen for you to make us supper.'

He dropped into a chair at the large dining-table squatting in one half of the room. The chunky piece of furniture doubled as his workbench during the weeks when it was too cold to work in the cellar. Taking up half of the scuffed tabletop today was a wide wooden cone, upturned and sitting in an iron cradle. Resting at the bottom of this shallow funnel were three iron balls, each smaller than Hooke's clenched fist. He scooped them out and set one after the other rolling around the wooden rim.

The growl of the iron on the wood blocked out his thoughts as his eyes followed their elliptical trajectories. Each ball would dip low and gain speed, then whip round the centre to climb the incline, never quite reaching as high as on its previous lap. He became lost in the repetitious motion, rolling ball after ball, wondering what would happen if there were no friction between the iron and the wood. Would the balls circulate endlessly like planets?

Grace reappeared wearing a simple shift with her brunette hair pinned rather brutally into the nape of her neck. In the lamplight she seemed to have less rouge on her cheeks now.

That's better, thought Hooke.

'You have new-fitted my room. Thank you,' she said.

Hooke rolled another iron sphere, and Grace went to the kitchen to prepare supper.

She returned almost an hour later, carrying two chipped bowls. Each was filled with a colourless gruel.

Hooke peered at the one she laid before him. 'Is it soup?'

'Potatoes and oysters.'

'From the pail near the window?'

'The oysters, yes.'

He pushed the bowl across the table. 'They were for an experiment next week on water temperature.'

'It's not my fault. I'm not used to this kind of work.'

'Oh no, you're too fine to cook for yourself, let alone others. Well, it's all you're good for now. You're nothing but a servant of your own making.'

'I won't spend the rest of my life in servitude,' she cried.

'Perhaps you should have thought of that before you let Sir Robert Holmes make a whore of you.'

She gasped. Her head dropped and she pressed her hands to her cheeks before speaking in a quiet voice. 'I thought that, of all my family, you might have forgiven me. We have a bond, remember?' She looked up. The hope in her face elicited a stab of hatred inside him. He knew exactly what she was trying to do, and the knowledge extinguished the provocative memories she was attempting to revive. 'Did you expect me to greet you with open arms? John is dead because of your shameful behaviour.' He stumbled from the table, seized the poker and stabbed at the embers. What on Earth had possessed him to spend this past fortnight fashioning new bedposts and remaking the chest of drawers in her room? What a fool . . .

'I have suffered enough,' she pleaded.

'My brother, my only brother. Gone because of you.' He glared over his shoulder at her.

Silent tears cut streams across her cheeks. 'Don't you think I would change things if I could? I never dreamt that Father would . . . '

'Go on, say it. Say it. Take his own life. Your father committed suicide because of you and your despicable behaviour. For all your London airs and your good looks, you're nothing but a foolish slut.'

Her eyes blazed. 'Indeed! I must have been utterly foolish to think that you would still love me.' She fled.

Hooke watched her disappear, an awkward mixture of guilt and satisfaction replacing the hatred that had swirled inside him. Had he not wanted her to cry, to show remorse? He thrust the poker back into the bucket, raising a sharp clatter. Yes, he knew, he had wanted those things. So, why did they now feel so wrong?

3
Woolsthorpe

Apples carpeted the orchard outside the manor-house, ripped from their branches by last night's gale. The wind had begun its journey over the North Sea, roused from its slumber at the urging of some great westward attraction. It had begun its blind tumble slowly, but by the time it reached the plains of Lincolnshire it was howling round the house's stone walls, piercing the gaps in the window-frames and invading the chimneys to blow soot across the hearths.

The maelstrom had roused Newton from his dreams. He lay for a moment trying to separate the real wailing outside from the one in his sleepy mind. In his dream he had been watching impassively as people screamed and shouted, their bodies tumbling in an avalanche of limbs towards molten lakes. He had been surrounded by tin furnaces and flasks of chemicals, the very things that had lain abandoned in Cambridge these past months. Shaking his head at the images, he had stumbled, sweating, from the bed and stared outside until the storm had abated.

Now, as he walked through the dewy grass, the air was still once more and suffused with the sweet tang of ripe apples. Around him, the servants were collecting the windfalls in large wicker baskets. He stooped to retrieve one of the fruits himself and rolled it from one hand to the other. If not for the fallen apples he could have dismissed the storm as part of the nightmare, the product of that dark place in his imagination he had promised to atone for had his mother been spared.

Years ago, during one of her haphazard visits, he had insisted that she walk with him through this orchard.

Pulling at her arm and ignoring her requests to know where they were going, he coaxed her further into the grounds. Choosing the most circuitous routes, he guided her into the furthest grazing-fields, where the sheep might look up with curiosity but no one else would notice them. With every step he had hoped that they would become so lost she would have to stay the night.

But she had turned him around with a soft stroke of his head. 'It's getting late,' she had said, even though it was early afternoon.

The memory jarred him back to the present. Those words were the last thing she had said before unconsciousness had taken her, and then death. *It's getting late*. What a hatefully dishonest way to announce her departure, then and now. He had not thought of it at the time, but now the phrase banged around his head.

He passed close to one of the farm hands. 'Send someone to examine the fences for damage,' he said.

The man touched his forehead in acknowledgement. 'Begging your pardon, Mr Newton, that's already been seen to. There's some little damage to attend to, but nothing serious.'

Newton pursed his lips. 'Very well. See to it that all is fixed by nightfall.'

'Yes, sir.'

Newton dropped the apple at the servant's feet and walked on.

He had to find a tenant farmer so that he could go home – yes, home. Cambridge was home: his books, his experiments, John Wickens. The yearning to return was becoming all-consuming. He and Wickens would start the experiments again. Perhaps the work would drive the troubling dreams from his head. The nightmares came almost every time he slept now and often lingered well into the next day. Something was coming, he could feel it. It was less than a shadow at the moment, but still perceptible. It was everywhere, a kind of latency, as if the whole world were waiting for something.

He looked at the servants gathering fruit.

Could they feel it too? He scanned the trees and above, to the islands of clouds in the sky. *What was coming?* An inner voice whispered a reply. He could not quite make it out, but it sounded like a single word: *apocalypse*.

4
London

Hooke climbed the tiny twist of staircase to the turret room, a fluttering sensation in his stomach. Over these last few days since Grace's arrival, they had settled into an uneasy rhythm in which each tried to avoid the other. They spoke in guarded terms, holding back all but the mildest hints of emotion. After each frosty encounter, something akin to remorse would churn inside him. He would have to fight the urge to comfort her, yet the moment she came into view, with those deep eyes and her perfect complexion, he felt repulsed.

Steeling himself, he knocked before opening the door to her room. The bed dominated the tiny space. The chest of drawers was covered with brushes, toiletries and a mirror. There was still a hint of the worked timber in the air, and the unmistakable smell of her particular rose-water.

He had not smelled that since . . .

She was sitting cross-legged on the bed, working with a bundle of mahogany-brown material, looking up at him.

'There is the Society meeting here this afternoon. Please remain out of sight.' He realised only after the words were out of his mouth that his intended command had come out as a plea.

'Will Edmond Halley attend?'

'No, he will not. He's in the South Atlantic, mapping stars from Saint Helena – not that his whereabouts are any concern of yours. You must stay in your quarters. No one is to know you're here.'

Her eyes dropped back to her work. 'You cannot pretend for ever that I'm not your Grace.'

Dangerous feelings bubbled inside him.

She pulled at a length of the fabric, splitting it down the seam. He recognised the dark folds as a pair of his breeches.

'What on earth are you doing?' He tried to grab the garment.

Lifting it out of his reach, she said, 'Gentlemen are wearing their breeches tighter these days, to allow for jackets being longer.'

'I don't have any long jackets.'

'Not yet.'

Hooke snorted. 'London is not the place you remember. Didn't the news of Sir Edmund Godfrey's murder reach the Isle of Wight?'

Grace shook her head. 'Never heard of him.'

'He was Justice of the Peace for Westminster until they found him on Primrose Hill, five days after he went missing, face down in a ditch, run through with his own sword.'

She looked unimpressed. 'There are always murders in London.'

'Not like this. You see, the wound hadn't bled. Sir Edmund must have been long dead when the sword was driven through his ribs. The doctors took the body and discovered a bruise around his neck.' With relish, he reached under his chin to grip his own neck. 'He'd been garrotted with a silk scarf. Later, the killers rearranged his neckcloth to hide the bruise, took his body to the ditch and skewered him to the ground.'

Grace stopped sewing. 'Why?'

'They say he'd taken an affidavit from Titus Oates . . . '

'Ah, now him I have heard of.'

'Don't interrupt. Oates said he had word about a Catholic plot to assassinate the King and return England to the Papists. He named more than five hundred people spread through the court, parliament, the regional manor houses – all of them secret Jesuits, just waiting for the Pope's order to begin the killings. You see, it wasn't just going to be the King; other prominent men were on the assassination list, too. Sir Edmund was murdered because he knew too much.'

Grace rolled her eyes. 'I hardly think we'll be in mortal danger going to a tailor to buy you a new jacket. I could do the final alterations like I used to. Besides, you go out all the time, it's only me who's stuck in here.'

Hooke huffed again. 'What do I care for fashion at my age?'

'You're not as old as you think.'

'I'm forty-four, that's old enough.'

'Exactly, you're not as old as you think.' She returned to her work.

Hooke stared at her. He managed a rather half-hearted 'Just stay in your room' before closing the door.

There was a chorus of polite greetings as the gentlemen straggled into Hooke's apartment, sank into the mismatched chairs that pinned the rugs to the floor and complained in unison of the cold weather. In years gone by the pre-meeting conversation had been about inventions and hypotheses; Hooke's fingers would tingle with the possibilities on offer and he would work late into the night, sometimes through to the next morning, in his fervour to manufacture the experiments needed to prove some point or other.

Springs, pulleys, balances: he knew how to make them all, relished making them as he imagined a master painter must enjoy being at the easel. These days, however, there were just the usual grumblings about the weather and the question of whether the Society's President would appear. Sir Joseph Williamson had been conspicuously absent these past weeks, no doubt consumed with his duties as Secretary of State, but still, an absentee President was not a good omen for any society.

Hooke's gaze was drawn to the grey metal cylinder on the hearthstones. Muted sounds of bubbling came from within its belly. In construction, the cylinder was not unlike an alchemist's furnace. At the base was a roaring firebox that heated an upper chamber. That morning, instead of chemicals, the contraption's owner had loaded it with a trivet and a dozen prepared pigeons. As the man had been fixing down the lid by turning a large screw assembly, he had stopped suddenly. '*Idiot!*' he scolded himself in a heavy French accent, pulling out a muslin packet of herbs and a bulb of garlic. He wafted them beneath his hooked nose. 'I nearly forgot these.'

'You do know this is only a demonstration, Monsieur Papin; an experiment,' said Hooke.

Papin's brow creased. 'Of course, but how can you cook a pigeon without garlic?'

Now the contraption shuddered and rumbled, filling the room with an admittedly delectable smell. Papin squatted in front of it, feeding more wood into the firebox. As the roar of the flames increased, so did the shaking of the metal body.

'Is that thing all right?' Hooke asked.

'Of course. Without the pressure, the bones will not be soft enough to chew. We will all eat well tonight, I think,' Papin said, flashing a grin. 'In the name of science, of course.'

Hooke returned to his secretary's ledger to record the names of the arriving Fellows.

The towering figure of Christopher Wren walked in, resplendent in a chestnut jacket trimmed with golden braid. Hooke noticed that the low waistline emphasised Wren's slim torso.

'Kit, I've been waiting for you. Gracechurch Street is at least two feet narrower than on my plans . . . '

'And good evening to you too, Robert,' said Wren genially. 'We changed the plans in committee – needed the extra space in the adjacent streets. Sorry, I thought you had been informed.'

'No, I was not . . . '

There was a commotion at the doorway. A wiry man with bulging eyes and a misaligned grey wig leaned on the doorframe, gasping from exertion.

'Mr Fisher, whatever is the matter?' asked Wren.

Hooke grabbed an armchair and swung it into position.

Fisher collapsed into the chair, the stiff fibres of his attire ballooning around him. 'I was followed on my way here tonight, I swear,' he said in a thin, frightened voice. 'A pair of men kept to the shadows behind me, speeding up and slowing down when I did. They were biding their time, I tell you. Lucky I was heading here and no further – they'd have had me for sure if I'd turned off the main street.'

The other Fellows crowded around him. A stout gentleman pulled out a short truncheon with a leather wrist-strap. 'I've taken to carrying this.'

'Mercy, Mr Banbury,' said Wren. 'What do you intend to do with that?'

'Defend myself. Any of those Catholic devils come near me . . . ' He slapped the weapon into his other gloved hand with a satisfying thwack.

'Come, do you really think there's evidence for these Catholic murder gangs?' asked Wren. 'It seems little more than rumours to me, all started by Oates.'

'The King must think there's something going on; he's housed Oates at Whitehall, presumably to protect him,' said Banbury.

'None of it makes sense to me,' said Wren.

'The papists are capable of anything,' said Papin, clapping his open hand to his chest. 'Why do you think I'm in England? We Huguenots are being driven from our mother country.'

'Well, I'm convinced. Especially after tonight. The danger's real,' wheezed Fisher.

Hooke fetched wine and the Fellows went back to their chairs. It became clear that their President was going to miss this meeting, too.

'I think we'd best start, Monsieur Papin. Don't you?' Hooke asked, lifting his quill to take minutes.

Papin bowed his head, regarded the Fellows and opened his arms with such vigour that he set his extravagant cuffs rippling. 'Distinguished Fellows of the Royal Society, it gives me the greatest of all pleasures to present you with my invention, the steam digester . . .'

'An invention based on my air-pump,' Hooke chipped in.

'Quite so,' said Papin. 'Where was I? Never again will the poor struggle to extract nourishment even from bones.'

There was an almighty ripping sound and the room filled with steam. Hooke instinctively closed his eyes as something hot spattered his face. When he opened them again, a boiling mist was rolling across the ceiling. The Fellows, many having jumped to their feet, were looking from one to another in varying degrees of shock. Fisher's wig – which had never been a good fit – was even more askew, and Wren's beautiful jacket was covered in pulverised pigeon.

Papin was on his knees, staring at the ruptured steam digester, covered from head to foot in dripping gravy.

'Tell me, Monsieur Papin, is this how they serve dinner in France these days?' grinned Banbury, licking his jelly-smothered fingers. 'I confess it is good, though.'

'I think we need to talk about fitting a steam valve, don't you, Monsieur Papin?' said Hooke, wiping his face.

Later, after the Fellows had drifted homewards, with Banbury insisting on escorting a tremulous Fisher, Grace appeared from her room. Her hair was bound up with strips of muslin, and she was clad in only a linen nightgown, apparently immune to the cool of the gathering evening. Hooke was on his knees, still cleaning bits of exploded pigeon from the rugs and floorboards. He tried to ignore the shadow of her figure underneath the fabric as she drew close. His buckled spine felt tighter than ever.

'Let me help you,' she said, crouching beside him.

She took the cloth from him, her nightdress gaping as she leaned forwards.

'For pity's sake, go and put something on.' Hooke struggled to his feet and slung one arm across the mantelpiece as if it were an old friend offering support. Blood thumped in his temples.

'Do you want something to eat? There's some cheese,' she said.

'No, I do not. The last meal you cooked kept me awake all night.'

'You never did sleep well.'

'My head fairly spun on the pillow.'

'There was only ever one remedy for your insomnia,' she said.

Hooke froze. He made to glare at her but found he could not meet her gaze. Instead he turned for the door, desperate to escape.

5
Cambridge

Huddled outside the weathered city walls and covered only in scrubby plants, the plague pits welcomed Newton home to Cambridge. Although fifteen years had passed since they had been filled, the mass graves still made him recoil. Inside the city gates was little better.

Whatever goodness may once have been found in the city's streets had fled as far as Newton could see, washed away on the sea of alcohol that flowed from the taverns and inns. The places to drink outnumbered all the other shops put together. Scabby children ran through the streets while drunken parents lolled against barrels and wasted the day. Market traders shouted their prices and foisted the poorest vegetables on those too intoxicated to see straight. Everywhere there was decay, and Newton shuddered even though his fascination kept him glued to the scenes.

They're damned, he thought, *every single one of them*.

From the carriage window Newton caught sight of the array of chimneys outlining Trinity College. Next, the four towers marking the corners of the gatehouse appeared. The magnificent sandy building rose from the squalid surroundings just as Newton towered over the beggars when he strode through the streets.

He rapped on the carriage roof, and the driver pulled the horses to a stop even though they were well short of the coaching inn. Newton wasted no time in disembarking and set off along the cobbles to where he could already see the college's unlikely herald waiting. Lurking in the shadow of the garden wall was a boy, dressed in sackcloth, with a shrivelled right arm. Newton guessed he was perhaps five or six. This seemed to be his patch. The steward would shoo him away periodically, but soon enough the urchin would be back. The boy nodded at Newton, who lifted his head to look at the college's gothic gates.

Choosing the smaller of the two wooden entrances, Newton still had to push with all his weight to make it open.

'Welcome back, Mr Newton,' said the chubby steward, offering the register for him to sign. 'Back for long, sir?'

'I sincerely hope so.'

Newton hastened to the quadrangle and paused to take a breath before slipping the latch of his door. 'Wickens?' he called.

The room was still.

'John Wickens, come out from wherever you're hiding – I'm home.'

There were neat piles of papers and books on the desk. The bookshelves were dust-free; the beds were freshly made. Newton's collection of prisms was perched on the top of the bookcase. The order made him uncomfortable, as if he had walked into someone else's room. He pushed a small pile of papers, sending them sprawling across the table, then he stepped through the hole that he and Wickens had made in the wall to provide access to the wooden shed where they performed their experiments.

All was neat in there too; the crucibles were stacked inside each other and looked as if someone had scrubbed them. The tin furnace sat in the centre of the room, lifeless and cold. The glass flasks were lined up in rows, and the alembics, with their pointed glass snouts, stood to attention in ranks.

It was too tidy. More than tidy, it was immaculate. Newton fought the urge to smash a piece of glassware.

Returning to the main room, he saw his favourite cushion had been plumped and perfectly positioned in his chair, like a cat basking in a sunny spot. He ran his fingers across the scarlet velvet so gently they left no impression.

On the table was a small collection of accumulated letters. Newton leafed through them, grimacing from time to time. Unless he was terribly mistaken, one was covered with Robert Hooke's handwriting.

The latch sounded. He dropped the letters and turned to the door.

'Isaac!' John Wickens was a slight man with delicate features and a smile every bit as mischievous as Newton remembered. The dark waves of his long hair curled into the hollow of his neck.

Newton embraced him. That was when he realised something was wrong. Wickens was tense, his usual ease gone from his body. Newton stepped back and waved his hand at the room. 'Did you stop working in my absence?'

'I have completed the notes for everything you did before you left. The notes are . . . ' He stopped when he saw the toppled pile of papers. 'Well, you seem to have found them already. Everything is up to date.'

'Excellent. I have new ideas. We must start at once.' Newton shrugged off his jacket.

Wickens turned away. 'I cannot help you.'

'Why ever not? We're close; I know it. I have new stirring patterns in mind. I think that seventeen clockwise rotations of the spatula, followed by a single turn widdershins–'

'I'm leaving the college.'

Newton stared at his companion. 'This is very inconvenient. Whom are you visiting? When will you be back?' Newton saw that Wickens's eyes were glistening.

'You don't understand, Isaac. There's a rectorship available at Stoke Edith in Monmouth. I want to get married. Start a family.'

Newton cupped one hand within the other and began to run his thumb across the ragged nails. 'When did this . . . this strange desire take hold of you?'

'It has always been in me to have children. I've made no secret of it.'

'Yes, but we've chummed for twenty years now. You haven't mentioned it recently. It's a whim.'

'It's no whim, Isaac.' Wickens turned to face him. 'Time's passing. You and I, we've had an extended springtime. I don't want to find myself suddenly in the grip of autumn.'

Newton's jaw began to tighten. 'Monmouth is far away.'

'I will always write.'

'Save your ink. I'll have no interest in hearing from you.'

'How can you mean that?' Wickens moved closer.

'Leave me, Wickens. You try my patience with your sentimentality.' Newton turned his back and held his breath, digging his fingernails into the palms of his hands.

There was a whisper of movement, then the creak of the door opening and the bang of it slamming shut. Newton continued to hold his breath.

6
London

It was the fifth of November, and London was in flames. Orange tongues twisted into the night from bonfires built on street crossings and patches of green. Cinders drifted upwards into the chilly air like freed souls racing to heaven. The people were supposedly commemorating God's deliverance of James I from the Catholic plot to blow up Parliament back in 1605, but to Edmond Halley something darker was permeating the revelry.

He tried to dismiss the thought as prejudice. He had been just ten years old when the Great Fire of London had raged across the city – a tragedy sparked by Catholic incendiaries, according to some. Despite the thirteen years since, unpleasant memories were still easily kindled. Halley recalled being bundled into the night, unsure yet excited by the atmosphere. He had been entrusted with a bundle of clothing and relied upon to walk alone while his mother carried his younger sister and gripped his little brother's hand. His father had led the way, laden with a chest of hastily packed possessions.

Swept along in a tide of people, Halley's nervousness had resolved into a sense of duty. Although there were noises all around – the occasional shout or sob, the barking of a dog or the whinny of a horse – he snatched only momentary glances in their direction. Mostly he concentrated on marching in step behind the broad expanse of his father's back, determined to keep up.

When the family squeezed past an empty cart being pushed towards them, into the city, its owner made a brusque offer to carry them to safety. Halley's father shook his head. A panic-stricken man carrying a frail old woman rushed in to ask the price.

'Twenty pounds to Moorfields.'

Two hundred times the normal price! thought the boy.

Above them there was a roar of thunder and orange sunbursts as first one roof then another caught fire. Cries of alarm went up as the flames took hold.

'Let's hurry up. No time to dawdle today.' Halley's father winked at him over his shoulder, and Halley found himself smiling back, insulated from the panic.

Ahead of them a stuttering Frenchman backed away from an irate mob. Halley caught a few words of his protestations and realised that he was pleading for his life. The mob were screaming that if the man was French then he must be Catholic, and that meant he was guilty of starting fires, but Halley had seen the way the wind was urging cinders through the night air and dropping them on the thatched roofs all around. New fires were sparking up everywhere. He tried to speak up but his small voice was lost in the baying of the crowd. A wild-eyed woman landed the first blow on the Frenchman's chin, and as the rest joined in Halley's mother ushered him onwards.

On the frosty slopes of Moorfields, thousands of people had camped out. Halley could still recall the horrified look of a young woman as she stared back at the dying city, her face illuminated by its angry light. She had conjured the strangest of feelings inside him – incomprehensible at the time. It had seemed wrong that a girl so beautiful should look so sad.

'It'll be all right,' he had said to his mother, reaching back to touch her thigh.

However, it had not been all right. In just a few years his mother, his brother and his sister would be gone for ever, each taken from him by one incomprehensible illness or another.

Shaking these dark thoughts from his head, Halley circled the fringes of a crowd gathered around a large bonfire. The heat on his face was in stark contrast to the frosty air that chilled the rest of his body.

Paying little attention to the path in front of him, he almost collided with a gigantic man backing out of a doorway.

'And I say that you, sir, are a charlatan!'

The voice stopped Halley in his tracks. *Robert Hooke?*

A hunched figure also stood in the doorway, rendered comically small beside the giant. Its crabby expression, narrowed eyes and sharp protruding nose were unmistakeable.

'Robert! It's me,' said Halley.

Hooke squinted.

'Edmond Halley. I'm back in England.'

Hooke's eyes widened. 'And not a moment too soon. This gentleman here was . . . '

The giant stepped forward. He was clad in heavy academic robes from head to toe. 'Good evening, Mr Halley, I am John Goad of Oxford.' He bowed with such a melodramatic flourish of his robes that, if not for his bulk, he might have taken flight.

'Don't start all that again,' said Hooke, flapping away the fabric.

'Mr Halley,' said Goad, 'I wish to address the Royal Society of London for Improving Natural Knowledge.'

'Then Mr Hooke is your man,' said Halley.

Goad and Hooke scowled at each other.

'He's the secretary and curator of experiments–'

'Also the Gresham Professor of Geometry and Surveyor to the City of London, but you, sir, are a peddler of nonsense,' said Hooke emphatically to Goad. 'The Royal Society only welcomes gentlemen inclined to the experimental sciences.'

'I practise the oldest of the sky arts.'

Halley sighed. *An astrologer . . .*

'We have rejected astrology as false,' said Hooke.

'But I can bring personal testimonies of the accuracy of my horoscopes–'

'One needs only to sing a madrigal every day for a fortnight, and lo! it will cure you of your fever,' said Hooke. 'Is that to be trusted? Of course not. Without a clear understanding of how one thing links to another, you can believe in nothing.'

'I am aware that my art is falling from favour, but God could not have created the heavens without purpose. And that purpose is the weather – why does it rain one day and shine the next? Why? Because of the planets! The air around us is the shoreline with the heavens. It must be influenced by the planetary positions.' Goad looked from Halley to Hooke and back again. 'The movement of the planets is the engine of our seasons – God's engine for raising our crops, providing the water and the sunlight needed for them to grow.'

Hooke tutted.

Goad smiled indulgently. 'I understand your scepticism, but consider this: Johannes Kepler, that paragon of astronomical achievement and the architect of the Sun-centred astronomy, was also a master astrologer.'

Halley said, 'Mr Goad, Kepler's laws of planetary motion can be tested telescopically and shown to be true. We reject the rest of his work. Astrology cannot be measured by any instrument that I know.'

'Do you similarly reject the Reverend Flamsteed? He cast a horoscope for the founding of the Greenwich observatory.'

'Have you seen that horoscope?' challenged Halley. 'I was there on the very day it was cast.'

Goad's face fell. 'You know the King's Astronomer?'

'I do, and written at the bottom of the page in John's own hand is *risum teneatis, amici – can you help laughing, friends*?

He cast his horoscope as tradition dictated, not because he believed in it.'

Goad opened his arms. 'My friend, astronomers can tell us where the stars are, but it is up to astrologers to interpret meaning. Without us, the study of the heavens is futile.'

'The study of the heavens will allow us to safely navigate the oceans. I hardly call that futile.' Halley pulled his jacket shut.

'We do not believe in strange planetary forces communicating themselves across space,' added Hooke. 'That is magic. We believe only what we can measure.'

Goad flung his hands in the air. 'Measurement! Always measurement!'

'The measurement of nature is the only sure way of leading us back to God. Yes, Mr Goad, when Adam looked into a drop of water, he saw the microscopic life within it just as plainly as I have seen it using my microscope. When he looked into the sky, he saw all the stars in heaven and in so doing, he saw God. But, during the Fall, man's senses crumbled, trapping us in these tiny brains with these limited senses. So now we have to build telescopes and microscopes to rediscover the knowledge of creation, and to rediscover God.'

Goad looked aghast at Hooke, who ploughed on. 'Once our investigations have revealed nature's laws, we will be left with a collection of unexplainable phenomena that must therefore require God's direct intervention.'

Goad spluttered, 'Smacks of blasphemy to me. I tell you this with great confidence: my system of astrological weather prognostication will prove more important than any of your efforts at weighing the air.'

'The term is *air pressure!*' cried Hooke, bunching his fists and resting them on his hips. 'And the barometer is an important scientific instrument!'

'Useless!' barked Goad, leaning forward.

Halley placed a hand on each man's chest. 'Enough! Let him go, Robert, there are those who will never understand.'

Gown fluttering, Goad took off into the night. The bonfire crackled. A hawker bellowed his sales pitch for roasted chestnuts. A series of sharp reports split the air, and the two men looked up to see the brilliant trails of rockets shooting through the sky and blossoming into colour high above. Their attention was drawn back to Earth by a noisy gang of young apprentices who swaggered past them, swearing and shouting at onlookers to make way. To the men's horror, the lads appeared to be dragging a body.

'Make way for his Holiness!' the gang shouted.

No, not a man, Halley realised, but an effigy dressed in the white robes and mitre of the Pope.

Egged on by the crowd, they hefted the straw man on to a pyre. Bawdy cheers erupted and drowned the roar of the flames.

'When did England become so mean-spirited?' said Halley.

'About the same time it dawned on people that the King was inching us back to Rome.'

'Do you really think Charles is a Catholic?'

'Of course. His brother has converted openly.'

'The Duke of York?'

Hooke nodded. 'Parliament has forced him to step down as Lord High Admiral.'

Halley could scarcely believe what he was hearing. Had England changed so much in the two years he had been away? 'But he's heir to the throne . . . '

'They're trying to exclude him from that, too. And, if they fail, we'll surely have a Catholic monarch – if we don't have one already . . . Tell me, how long have you been back?'

Halley inhaled the smoky night air. 'Some weeks, well, maybe a month or so.'

'And only now you want to see me?'

'I've been busy making the final calculations for the star chart. I have to present it to the King next week, and it has to be perfect.'

'And you couldn't even drop in to say hello?' Hooke eyed him expectantly.

'Robert, I'm not the only thing that's back. So is the comet.'

'What?'

'The comet of last month has returned.'

'I haven't seen it.'

'It's not in the evening sky any more. It's appearing before dawn. It's as if it fell towards the Sun, swung round behind it and now travels outwards into space again.'

Hooke's eyes glittered in the firelight. 'Can we see it?'

'If we wait the night out . . .'

It was nearly dawn before the bonfires died away into tiny, smoky columns dotting the city's skyline. The dampness in the air meant that it would be a race to see the comet before the soot combined with the morning mist to choke the city in a filthy brown fog.

Halley scanned the horizon from the cramped roof platform atop Gresham College that Hooke had commissioned for his astronomical work. Hooke stifled a yawn and pulled out a handkerchief from his leather money pouch to wipe the lens and tube of the long, thin telescope, preparing it for use.

Then Halley saw it, just a glimpse at first – a ghostly fan of light, no bigger than a thumb's width at arm's length. It was hanging above the rooftops, so faint that it danced in and out of visibility.

'There it is,' said Halley, pointing into the sky.

Hooke jostled in beside him and peered into the darkness. 'You're imagining it . . . ' He fell silent.

Halley slid away to swing the telescope into position and duck his head to the eyepiece. *Beautiful!* The comet's tail reminded him of the long hairs that trailed a galloping horse, frozen in a portrait. He traced the threads of ethereal light to the head of the comet, where a small jewel glittered. 'Come and look.'

Hooke bent to the eyepiece, knocking the tripod and forcing Halley to realign things. 'Well, I'll be . . .'

'Flamsteed thinks that it must be magnetic and that it's being repelled by the magnetism of the Sun,' said Halley.

'Impossible! If you melt a magnet it loses its magnetism, and the Sun must surely be molten to be so hot. Even Newton agrees with me on this.'

'You talk to Newton? I must have sailed back to some strange country and mistaken it for England.'

Hooke looked round from the eyepiece. 'Not even I can hold a grudge for ever. We've been exchanging letters.'

'The controversy over the origin of colours is forgotten?' Halley moved closer, unable to keep the disbelief from his voice.

'Of course. I asked him if he could use his new mathematics to calculate why the planets stay in orbit.'

'To prove why Kepler's laws work?'

'Precisely.'

Halley was impressed. 'I admit I don't know much about his new mathematics. What's it called? Fluxions?'

'Fluxions and fluents,' Hooke said. 'No one knows much about it. He prefers to keep the method a secret, though I believe there are some papers lodged in the Society's archives, and Collins once told me that Newton had exchanged something with Leibniz in Hanover – apparently he was thinking along similar lines.'

'Leibniz was?'

'Yes. Both working on a way to calculate the rates of change of moving quantities . . . but you're distracting me. Newton isn't really interested in Kepler's laws any more. He says he's abandoned natural philosophy altogether, *and* he made a number of mistakes in his letters that I corrected for him.'

Halley registered the satisfaction in Hooke's voice. 'So, what is Newton doing these days?'

'Alchemy,' said Hooke.

'But that's illegal. You must be mistaken.'

'From what I hear, he hardly ever leaves his furnace.'

'I didn't have Newton down as a dabbler,' Halley said.

'He's consumed by some notion of concocting the Philosopher's Stone.' Hooke suddenly looked impish. 'Perhaps if he's thrown in gaol, he'll have some time to perform our calculations for us.'

'Perish the thought,' Halley said, making Hooke chuckle.

The unexpected sound of someone climbing the rickety staircase caught Halley's ears. He turned and locked eyes with Grace.

'Mr Halley,' she said, her eyes smiling at him even if her mouth stayed level.

'Mistress Hooke, I'm delighted to see you again.'

'Will you be staying for breakfast?'

'He's not staying. He's just leaving. Off you go.' Hooke bustled round, barring her entry to the platform.

The woman allowed her gaze to linger on Halley even as she complied with her uncle's wishes and descended the staircase. Her image seemed to linger in the air where she had been standing.

'Grace has grown somewhat since last I saw her,' said Halley.

'Keep well away from her, Edmond.'

Hooke's tone was sharp. The young astronomer mumbled, not entirely with veracity, 'I remark on her as I would a spring flower.'

Hooke folded his arms. 'If you go now, you can still get a few hours' sleep before the day starts properly.'

'As you wish.' Halley knew better than to argue when Hooke started behaving like this. He made his way to the stairs; the handrail was slick with dew. Two steps down, he stopped. 'Robert, just one thing. What you said to Goad, about finding God in the things we cannot understand. Did you really mean that?'

Hooke gave him an exasperated look. 'Of course.'

'But what happens if we can explain everything? I mean: what if all our experimenting, all our observations and all our measuring give us a set of mathematical laws that can explain everything? What will we need God for then?'

'That will never happen.'

But what if it does? Halley thought. *What if it does happen?*

7

Halley awoke with a start at the sound of a horse and carriage outside his bedroom window. After chipping away enough of the ice from the inside of the pane to see the vehicle disappearing around the street corner he let his forehead drop to the frigid glass.

Tugging on yesterday's clothes, he bounded downstairs and pulled open the front door. A blast of icy air reminded him to reach for his jacket and shoes. *Lord knows what would be said if I turned up at church wearing only a nightshirt,* he thought. Despite the throbbing of his head, he forced himself to run after the carriage, barging his way through the Sunday-morning pedestrians. He knew he was behaving abominably, but he had bigger concerns: he had promised his father he would attend church.

Thankfully the bells were still pealing and the congregation were still taking their seats when he arrived, panting and red-cheeked. He tucked in his shirt and smoothed some of the tangles from his hair. He found the correct pew and, apologising repeatedly to the worshippers he forced to stand, slid in beside the fleshy bulk of his father.

'We'd nearly given up on you,' said the old man, not turning to look at him.

'Someone could have sent the maid to wake me.'

'We thought you'd gone beyond needing a nanny,' came a sour voice from beyond his father.

Joane. Halley cursed her. She had arrived in his life uninvited five years earlier, a mousy woman with squinty eyes and freckled cheeks. Still struggling to come to terms with his mother's death, he had returned from the first term at

Oxford to find Joane in her place at the Christmas table. By the time he had come home for Easter, she was the second Mrs Halley. He soon began to dream of sea voyages and overseas adventures.

'Please tell the maid to salt the steps tomorrow before my father has an accident,' he said tartly.

With a smirk she returned to snooping on the rest of the congregation.

The vicar appeared and the congregation settled into respectful silence and stood up. He was a solid figure with pale brown hair that skimmed his collar, but the feature Halley noticed the most was the wild look in his eye; it would not have been out of place on a poacher.

'The dead shall rise again,' he thundered as he launched into his sermon. 'When the time is right, those who have lain in the ground these many years will rise and join those of us still living, and each of us will kneel before the Lord and justify our lives and actions. It's not the moment of death that we have to fear, but the inevitable judgement that will follow. That is why we live as God dictated through our Saviour, Jesus Christ. The good will die along with the wicked, but when Christ returns and judgement is made, only the good will be truly saved and raised into Heaven, where we shall spend eternity basking in the glory of God.'

Halley glanced at those around him, all transfixed by the vicar's words, fervour written on their faces. Even his father's features had softened, rolling away the years. *What was it they felt?* Confusion twisted within Halley. He stared past the congregation, at the walls and windows, and tried to see more than just stone blocks and leaded glass.

On his voyage to Saint Helena, the ship had been becalmed. Roasting under the fearsome sun, he had prayed like the rest of them for a breath of wind. He had searched inside for his faith, looked out to see some evidence of the Almighty. Yet he found only questions. *What would happen when Mother met Joane in Heaven? How would Father choose? How would everyone who ever lived fit into Heaven? Why were there no answers? Why didn't people seem to mind that there were no answers?*

After the service had ended and the chit-chat was over with, Halley felt morose. In the homeward carriage, he faced his father and Joane in silence. Halley's mood was not helped by the way his father looked these days. Apart from those few moments in the church, the upright bearing had been replaced with a stoop, the strong body with a portly roll, and the vigour with a couple of noticeably baggy eyes.

'Tomorrow we must get you some new clothes,' said his father. 'We cannot have you meeting the King looking as though you've spent all night on the street.'

Halley looked down; the dark weave of his jacket and his once-white shirt cuffs were particularly grimy.

'The King,' Joane huffed dismissively.

Halley shot her a look.

'Now then, you two . . .' said his father to no avail.

'Perhaps you will finally admit the value of my work?' asked Halley.

'Oh, I'm supposed to be impressed, am I?' she said, meeting his gaze.

'Have you never wondered why things are as they are? The world is not a secret place; it's open for our investigation,' said Halley.

'What good will such knowledge do us? Will it protect us from penury? Cure us of disease? You behave like a boy, concerned with boyish things. When will you grow up and stop expecting the rest of us to subsidise your foolish fancies?'

Halley noticed the hasty way his father looked out of the carriage window.

'So it was you who persuaded Father to reduce my allowance.' Triumph lit her face, stoking Halley's annoyance.

'One day I shall inherit my father's estate. Then you'll see how carefully I conduct my business, and I can assure you that I will not be in the habit of handing out loans to your acquaintances.' Halley saw the words strike home.

'And just where does all your money go?' she challenged him.

'Enough, you two. We're a family,' sighed Halley's father. 'But, Edmond, it is time to start thinking of a wife.'

'Yes, someone to calm you down,' said Joane.

'I don't need calming down.'

'What about Mary Tooke? You like her.'

'Father, she's a child.'

'She's older than you think. She'll be out soon enough, it's never too early to start these things.'

'But I'm to tour the observatories of Europe.'

'More of your father's money wasted,' said Joane.

'Reputation-building, my love. Edmond must show off his star catalogue, otherwise the effort will be wasted.'

Joanne huffed and pursed her lips.

'You go off as planned,' said Edmond the elder magnanimously. 'I'll start the negotiations regarding Mary. I'll write when there's an agreement.'

Halley decided the best course of action was to ignore it all. That way, like so many of his father's sudden fixations, the whole idea might just drift away as mist on an autumn morning.

'Yes,' piped up Joane, 'marriage will put an end to your debauchery.'

'I'm an astronomer.' Halley crossed his arms. 'I do not debauch women.' *Well, not unless they want me to,* he mused.

The Sun had not yet risen when, a few days later, Halley worked himself into his new clothes. He held a polished metal plate at arm's length, craning his neck, wishing that someone would invent a full-length mirror. The buff colour and the yellow trims went against the season but ensured that he would stand out. He raised his head to the window, where the rushlight gave a faint reflection on the panelled glass.

He liked the way the curve of his stockinged legs disappeared into the flared hem of the jacket, making his breeches hardly visible at all. He brushed his hair with care, arranging it over his shoulders and making sure that none of it was trapped inside the white lace stock around his neck. He lifted the rolled star chart, as thick and as long as his arm, from its place on the dresser. With a final nod towards his reflection in the window, he headed out of the room.

His father was perched on the box seat in the hallway. Dressed in the informal blue uniform of the Yeoman Guards – the red livery being reserved for days when the King was in attendance – he was struggling with his footwear.

'Let me help you, Father.'

'Can no man in London make shoes to fit my feet?'

'Perhaps if you wore fewer stockings.'

'I have to wear three pairs to ease the infernal rubbing of the leather. And, besides, it's winter.'

'Summer doesn't change the habit, from my memory.' Placing the star chart to one side, Halley managed to squeeze his father's foot into one of the offending leather vessels. 'You're in a bad mood because you have to attend to your duties at the Tower. You've grown to think of it as a club that you're free to attend at will, rather than as your duty.' He finished lacing the shoe and went to work on the second.

His father grumbled and tightened his belt, smoothed the straining blue fabric over the curve of his stomach.

'There,' said Halley, 'all done. Buy yourself some new shoes on the way home.'

'Perhaps I would, if I weren't owed over a hundred pounds by one of your mother's friends.'

Today was not the morning to remind him that Joane was his stepmother. 'You're referring to Cleeter?' asked Halley.

His father nodded. 'Then there's the Rector of Sawtry. He's always behind on the rent – a rector!'

'Why not let me help with your affairs? Your memory is not so good these days.'

'Have I not just demonstrated how good it is? What more do I need to remember? Besides, you're busy with your studies.'

'Not too busy to help my father. You run the soapworks, you administer the properties – it's too much.'

Silence hung between them.

'I was once like you,' said Halley's father before reaching for his periwig. He swung the dusty object on to his head, flattening the remaining silver wisps, and lifted himself to his feet. 'We must go.'

With a shrug, Halley retrieved the chart and unlatched the front door, allowing in the winter air.

'I'm very proud of you, son,' his father said, gesturing towards the rolled paper Halley was cradling. 'You'll do me proud again today.'

Halley looked into the old man's rheumy eyes. 'I will.'

After dropping off his father at the Tower, the carriage crossed London and rumbled into the greenery, heading for the King's dwelling. Fronted by the open spaces of St James's Park, and backed by the Thames, Whitehall was more than just a palace; it was almost a town in itself.

Extended and enlarged over the previous two centuries, it flaunted its mismatched architecture. Red bricks, grey slates, white stones – all played a part in this sprawling construction. It looked more like a collection of individual buildings than a single continuous dwelling. Its fifteen hundred rooms rendered it the largest palace in Europe, surpassing even the

Vatican, which was entirely as it should be for the home of an English monarch, according to general opinion.

Halley had been free of nerves for most of the journey, but now that the towers of the palace gatehouse were in view he felt his stomach flutter. Involuntarily, his hand tightened on the star chart.

Red-coated guards directed the driver to a courtyard, where Halley was greeted by a sallow official who introduced himself as Winslow. Without another word the official led Halley inside, where it seemed a redcoat was stationed every few steps. They served to heighten Halley's anxiety.

Perhaps the country really is in as much peril as Hooke and others are saying, he thought.

With no conversation to distract him, Halley found himself studying the cut and stitching of Winslow's clothing. It might have been drab in colour, but it was expensively tailored. Whoever this man was, Halley decided, he was important.

They arrived at a high-ceilinged ante-room with extravagant coving and white panelled doors. His guide swung on his heels and held out his hand. 'The chart, if you please?'

'I thought I would have the chance to address the King before showing him the chart.'

'You cannot march into the King's presence with something that has not been fully inspected.'

Like a scolded child, Halley handed it over. Winslow disappeared through a set of double doors. Halley drifted to the arched windows and looked out over a yard where a stable lad was grooming a horse. He turned back at the sound of the double doors opening inwards.

'You may approach His Majesty,' said Winslow.

Halley had expected the King to be ahead of him, enthroned and waiting, but the wooden seat was empty. The bow he had been practising stiffened inside him, and he ended up walking uncertainly into the room.

The tall frame of Charles II was bent over a table near a bay window. He looked around at the sound of footsteps. 'Edmond Halley, come in. We don't stand on ceremony here.'

The chart was unfurled across the table.

'This is a fine work.'

'Thank you, Your Majesty. It's still a little crude compared to the real thing I intend to publish. It's accurate, but it lacks the artistic finesse I envisage.'

'I'm told that you left your degree in Oxford to pursue this work.'

'I felt it was more urgent than what I was being taught.'

Charles indicated the chart, on which a large circular frame represented the horizon and a rich scattering of black spots marked the positions of the stars Halley had observed. The astronomer had sketched in a few lines to indicate constellations, both new and old. Familiar ones nudged in from the top and followed the sweep of the zodiac – Capricornus, Sagittarius and Scorpio – but most of the others made alien shapes. 'I'm intrigued by this grouping here,' said the King.

Halley knew which one the King was referring to without looking. He had made the pen stroke just last night while putting the finishing touches to the chart, marking out a straight alignment of stars and labelling it *Robur Carolinum*, Charles's Oak.

'I fear you think me presumptuous, sir. I had thought to speak to you before I showed you my map.'

'On the contrary, I think you're ambitious.'

Halley looked at the King's face for the first time since edging into the room. Under a mountainous black periwig, the King retained his swarthy appearance. Although his cheeks had puffed with age and the flesh around his eyes had darkened, he was still a handsome man.

Halley spoke. 'As the ancients placed their stories in the skies, so I took the liberty of placing your own story there too. What could be more inspirational for our mariners a long way from home than to look up into the skies and be reminded of the tree that sheltered you from the Roundheads? With your permission, I would like to dedicate the catalogue in your honour, and in recognition of the help you provided in your instructions to the East India Company to provide me with passage.'

A curious smile lit up the King's face. 'I will allow this because you, Mr Halley, are something of a wonder: a gentleman, an able navigator and an accomplished astronomer. Yet you don't look old enough to grow a full beard. Now, the first makes you acceptable, the second makes you an asset, and the third – ah yes, the third – that makes you invaluable.'

Halley's mind started tumbling.

'I know of the so-called invisible college spread across Europe,' continued Charles. 'All you astronomers and natural philosophers writing to each other even when your parent countries are at war. This new investigation of nature seems to know no boundaries.'

'I am English to the core, my liege.'

'Indeed you are. And, as an astronomer of note, you will now be welcomed in countries across Europe regardless of their political or religious persuasion. You have already received a number of invitations, I believe: Danzig, Paris, Rome.'

'Your Majesty is well-informed.'

'Winslow does his best.'

Halley glanced over to where the scrawny man hovered.

'Walk with me,' the King said to Halley, striding off towards a set of doors deeper in the room. Guards stationed either side opened them at his approach and then fell into step some half-dozen paces behind. Winslow drifted along at the rear, feigning incuriosity. Halley made sure his shoulder remained behind the line of the King's as they traversed a white-tiled corridor.

'What if I were to instruct Oxford to grant you your degree in recognition of your star chart? Such a strategic work of national importance should be acknowledged. Whoever controls the oceans controls the world, and to control the ocean we need accurate star charts.'

'I should be greatly in your debt, sir.'

'Do you know, it's a shame you weren't born a few years earlier. I might have named you my astronomer instead of Flamsteed. He's so tardy with his work.'

Halley felt a pang. To be the King's Astronomer would indeed have been a prize; he did not know what to say.

'I have a modest elaboratorium here in Whitehall,' continued the King. 'Alas, affairs of state conspire to keep me ever further from it, but one similarity I have noticed between natural philosophy and affairs of state is that, to be successful, both require . . . information. If I were to help you again – with finance for your coming trip into Europe – perhaps you would make a mental note to tell me upon your return if there was anything you noticed while you were away. Perhaps you might overhear certain things of value to me in my handling of the foreign powers. Maybe you will see some fortifications on your travels that it would be advantageous for us to know about.' The King paused a moment. 'Do you think you could do that?'

Halley felt a stone lodge in his chest. 'Espionage, sir?'

'Not at all, I'm just asking you to be vigilant. You see, it's seldom as straightforward as a matter of friends and enemies. There are balances to be struck, evidences and outcomes to be weighed. That's where diplomacy comes in, and information is the key to good diplomacy. If I have one piece of advice for you, my young friend, it is this: trust your own judgement and always be wary of the promises of wise men. But the promises of powerful men, they are something different altogether and should never be mistaken.'

Halley's mind was working furiously to extract the meaning from the King's words. Something about his emphasis on the word 'mistaken' made Halley suspect he had just been threatened.

The King turned to face him. 'Are you with me, Mr Halley?'

Halley turned too. He took what he hoped was a discreet deep breath. 'Yes, Your Majesty.'

Out of the corner of his eye he saw an ugly expression split Winslow's lips. It took a moment to realise it was a smile.

8

Black velvet with red piping, golden buttons and extravagantly turned cuffs, Hooke's new jacket was without doubt an exquisite piece of tailoring, though he generally found fine metalwork more worthy of admiration.

'Go on then, try it on,' said Grace, dancing from one foot to the other.

Self-consciously, Hooke slid his arms into the jacket. As he fastened the buttons, she crouched and tightened the lacings at the knees of his breeches, then she brushed her hands across his shoulders, caressing the sweep of his arms and the contours of his back in a single flowing movement.

'Is it as you had hoped?' he asked.

'Oh yes,' she said, stepping back to appraise him head to foot. Her radiant expression made him glow inside. 'Do you not feel the difference?'

He indulged her with a smile. 'If it makes you happy, it makes me happy.'

'You'll see; people will treat you differently now. And what about me?' She pirouetted. A perfect emerald-green dress emphasised her waist and an elegant shawl encircled her shoulders.

'You're beautiful.' Suddenly emotional, he turned and retrieved some letters from the little table near the front door. Before he knew what was happening, her lips pressed against his cheek, just a hair's breadth from his mouth.

'Before you go: remember, we saw those posters advertising dancing lessons? I would like to go to them.'

'Now is not the time for us to be discussing this,' he stammered.

'I cannot stay inside for ever.'

'But you do go out – I take you out. Come with me now to the Letter Office,' he said.

'That's not what I mean.'

Hooke looked into her eyes. 'I'm an embarrassment. Is that what this new jacket is all about?'

'No. It's just that . . .'

'Just what?'

Grace went to the misty window, where twilight was transforming the city outside into dark shapes.

'Just what?' insisted Hooke.

'There are some things I would like to do on my own. And I could shop for us. Why should you do that, when you are so busy? You hardly sleep at all these days. I hear you pacing in the middle of the night.'

Hooke felt embarrassed that she was aware of his insomnia. The other night, he had been so possessed of nocturnal energy that he had taken a saw to the beams in the cellar, cutting a niche so that he could move the big display cabinet into a more accessible position.

'Why are you afraid of letting me out on my own?' she asked.

'I know the thoughts that dwell in your head. The indiscriminate way you apply yourself. Remember last time.'

Her little nostrils flared. 'I have been nothing but saintly since my return. Am I to be given no credit for trying to mend my ways?'

Hooke pulled a face and shuffled the letters.

'No, wait.' There was a sharpness to Grace's voice now. 'That's not the problem, is it? This is not about what I might do with others; it's about what I don't do with you any more.'

Flames of passion and shame burnt inside Hooke. 'I can always send you back to the Isle of Wight.'

'You wish that we'd taken up where we left off, don't you? You and I.'

'Never speak of that again. It was a dream, nothing else.' With that, he strode outside and into the gathering night.

He was grateful for the coat almost at once. As well as being warmer, the longer length did make him feel as if he were walking more upright. He noticed similarly flattering designs on other gentlemen and wondered whether he looked as affluent as they did.

There was a small queue in the General Letter Office when Hooke arrived. Once the staff had stamped his letters with the date and disappeared with the folded sheets into the back of the bureau, he returned to the streets. Though he had only been in there a few minutes, the light had gone. Rather than return to Gresham and Grace's rebellious mood, Hooke's mind filled with the lure of some hot coffee and gossip. He turned his back on the college and headed towards the dark alleyways of the coffee-shops.

Why does she always have to be so ungrateful? Does she not realise the humiliation I risk by taking her in? I protect her and what precious little is left of her reputation, yet she doesn't . . .

There was a sharp crack on the back of his neck. The impact ricocheted along his shoulders and up into his head. He sensed himself falling and felt the hard slap of the freezing earth on his forehead. A wretched smell clung to the dirt, yet Hooke was unable to summon his body into any form of response. In the pale grey moonlight, black shapes knelt over him. Hands dug through his clothing. There was the sound of ripping.

'Hardly nothing,' a voice said.

The shadows moved, and Hooke closed his eyes, expecting more blows. There were some sharp touches as the few coins he was carrying were contemptuously thrown into his face. Then the footsteps faded away and silence followed.

After some indeterminate time, feeling started to return to his body, and with it, pain. His head throbbed. He touched the tender spot on the back of his head, then his forehead, and traced blood from the wound down his left cheek. Shaking violently, he struggled to his feet and looked around. The road did not look at all familiar. He must have taken a wrong turn. Fifty yards away, he could just make out people walking past the street entrance.

Bishopsgate?

He hobbled to the main road, where city gentlemen skipped smartly across the way as he emerged from the darkness. Hooke wanted to remonstrate that he was a victim, not a vagabond, but every time he tried to catch somebody's eye they quickened their pace.

Strength ebbing away, he just made it back to the apartment. He lurched through the door and blackness overwhelmed him.

When his eyes opened this time, full green skirts were racing towards him across an expanse of floorboards. He could feel the vibration of Grace's footsteps jarring his brain.

She lifted him to a sitting position, and his sight cleared a little.

'Let's get you up,' she said, helping him into a chair. 'What happened?'

'Catholics,' he croaked.

Her footfalls receded, then returned.

'Drink this,' said Grace.

He fixed both hands around the proffered goblet and lifted it to his lips. The bite of the red wine in the back of his throat was comforting. He drained the goblet. The still-fuzzy outline of Grace knelt in front of him, rinsing a cloth in a bowl of caustic-smelling water before lifting it to dab his forehead. His eyesight returned to normal at the sharp sting.

There were wet marks on his coat, and twists of blood curled in the washbowl.

The goblet slipped from his fingers to clatter on the floor. He rocked back and forth as tears fell from his eyes. 'I'm sorry,' he whispered to Grace, repeating the words over and over again.

She slid her fingers into his palms.

'I don't mean to hurt you. I want to protect you.' He pulled her close, leaving her no option but to shuffle on her knees between his splayed legs. He anchored himself to her, repeating his apology between sobs.

She wrapped her own arms around him, stroking his hair and drawing him closer still. 'It's all right, I'm here.'

Her warmth seeped into him. The swell of her breasts against his chest reminded him of the last time they had been this close. He tried to banish the memory but it was too powerful. It had been gnawing at him ever since he had agreed to take her in again, shameful but stimulating. He no longer had the strength to fight it. His mind returned to three years before.

It had been another icy winter, less than a year after she first arrived. The Thames had frozen, and no matter how many blankets he piled on his bed, nothing could keep out the chill. One night, with chattering teeth, he had crept up to her room and pushed at the door.

'Are you asleep?' he whispered.

'No, it's too cold.'

'Let us snuggle together, combine our warmth.'

She rose without a word and followed him to his room. He got into bed and lifted the covers for her. She slipped inside, turning away to lie on her side. He took it as invitation, and she did not complain as he stretched along the curve of her back, drawing himself into full contact with her.

At first her warmth was a comfort and sleep beckoned, but as more of her warmth entered him, the reality of their embrace took hold. His erection pressed through the thin fabric of their nightwear, against the firm flesh of her buttocks. He knew she was awake by the tension he could now feel in her body.

First he pretended to be asleep, then he tried to ease himself away, but the slight movement sent a frisson through his whole being.

There was no Moon that night. Cocooned in the darkness, her soft fragrance grew overwhelming. He moved again and another shiver rippled through him. A small breath escaped him.

The covers moved, and he felt her hand creep across his belly. She rolled on to her back and gripped him. The freezing air on his face accentuated the warmth of her hand, and within moments he shuddered with release.

'There,' she said, 'that'll stop you fidgeting.'

Next morning he said nothing, even though it was all he could think of. They went about their daily rituals and the milder weather returned them to their separate beds.

He had almost convinced himself that it had all been a dream, when the cold weather returned and in shamefaced desperation he invited her to his bed again. The scene played out as before, and the pattern was set.

Befuddled by the pleasure she seemed so willing to dispense, he persuaded himself that her nocturnal ministrations were little more than innocent dreams come to life, but then came the day he returned to Gresham and heard raucous laughter from her room.

The sight of a greasy-haired tanner's lad, his stained fingers probing her soft flesh, sent Hooke into a violent rage. He shouted, screamed and broke furniture that he still missed today. The boy had run for his life despite being twice Hooke's size.

The next day, Hooke began the arrangements to send Grace back across the Solent, thinking she could be weaned off her lascivious behaviour. Instead, her exquisite looks and London ways had attracted the attentions of Sir Robert Holmes, the island's governor. What began as a public flirtation became a worrying courtship and ultimately a careless conception. Holmes rejected Grace on hearing the news, but agreed to take the child. However, this final act of benevolence came too late to save Grace's father, who, humiliated, had hanged himself. With no one willing to forgive Grace her sin of motherhood, Hooke had agreed to take her back.

As she tended his wounds, the wine and the memories bored into him. Her proximity inflamed him. His breathing started to quicken and he dared not look in her eyes. She was looking at him, understanding written all over her face. His body pulsed, the pain displaced with desire. Wordlessly, she found the catches on his breeches and worked her hands into his clothing.

Hooke surrendered.

Afterwards, as if it had been a part of the medical treatment, she washed her hands and smeared witch-hazel on the purple bruise she had revealed on his left hip. Then she fastened up his clothes again, dressed his head wound and peeled off the damaged coat, inspecting the tears. 'Easily fixed,' she said.

Hooke wobbled to his feet and dragged the chair closer to the fire. Grace fetched more wine. They sat in the flickering light, Hooke in the chair, Grace on the rug at his feet with her head resting on his thigh.

'What were your father's last words to you?' he asked.

'I don't remember; they were words of insignificance, spoken before he knew I was carrying the baby. He never came to see me after my confinement began. No one did, apart from Mother and the maid.'

He reached down and rested his hand on her shoulder, touching her as gently as he would a piece of glass apparatus. 'You're safe now.'

'After your ordeal tonight, I think I should be saying that to you, *Uncle*.'

9
Danzig, Polish-Lithuanian Commonwealth

Halley was more than ready for a drink. He lifted the pewter tankard from the maid's tray and raised it to his lips.

'Ah . . . I was to say a toast,' said an accented voice.

Halley froze.

Johannes Hevelius – Jan, as he had insisted on being called at the river port – smiled. His fleshy bottom lip was almost concealed by a full black moustache that spread to each cheek. 'As my father said, nothing should stand between a gentleman and his beer.' He gulped a great mouthful and smacked his lips. 'I cannot express how delighted I am to have you here, Mr Halley.'

'Edmond, please. It's been a long journey, but it's a pleasure for me to be here.'

'I think they send you to convert me to telescopes.' Jan's eyes gleamed.

'Your accuracy without them has been raising a few eyebrows back in London.'

There was a rustle of clothing at the door. A motherly figure, wide and round as a tree trunk, swathed in layers of olive-green fabric secured by saffron-yellow ties, filled the doorway.

Jan extended his free hand. 'May I present my observing assistant, calculator and wife, Elisabeth.'

'Delighted to meet you, Madame Hevelius,' said Halley as she entered the room.

She returned the greeting, her voice softer than Halley expected. He looked into her plump face. She had young, unwrinkled eyes and her chestnut hair showed little sign of grey. He glanced at Jan.

'Yes, Edmond, she is much younger than me. By thirty-six years! And, to save you guessing, I'm now sixty-eight.'

Halley opened his mouth to apologise, but Jan's eyes were creased in mirth. In fact, he looked rather pleased with himself. 'I'm only old if you count the years since I was born,' he laughed, raising his beer again.

Halley did likewise, unable to refrain from contemplating his host now that ages had been mentioned. Only Jan's steel-grey hair betrayed his age. He was tall, still comparatively lean, and appeared strong. His waxy skin had relaxed its grip on his flesh but his features remained well-defined. Halley found it difficult to believe that he was talking to a man older than his own father.

'I know what I have that you will like,' said Jan, hurrying from the room.

'Tell me, Mr Halley, what are dresses like in England these days?' asked Elisabeth.

Halley shrugged. 'Generally they contain less fabric–'

He broke off, cursing himself for the double meaning. Thankfully, Elisabeth just laughed.

She stepped closer. 'You are to be greatly admired, Mr Halley. A man of your youth, yet already so distinguished by achievement.'

Her praise did not ease Halley's embarrassment.

'How old are you now, Edmond?'

'Twenty-three.'

Elisabeth's eyes were the deepest brown that Halley could remember seeing.

'I was already married at that age.'

'My father is doing his best.'

'Jan took me to the altar when I was sixteen. By twenty-three I was helping him every night in the observatory, but I have not sailed to distant shores, nor completed a star chart. Tell me, what's it like on foreign lands?'

'On Saint Helena the light is so much brighter, the day so much hotter, the air more invigorating . . .'

She watched him intently as he talked.

'I remember one couple who arrived – a gentleman and his wife. He was fifty-five, she fifty-two. I know because I was bold enough to ask them after an incident that occurred shortly before I left the island. Shall I tell you? Well, they were to settle on the island to live out the ends of their lives in peace. The air had such a beneficial effect on them both that word was soon everywhere that . . . that . . .' He felt himself colour. *Why did I start this particular story?* 'She, em, because of the restorative qualities of the air, you understand . . . she found herself with child, where none had been possible in England.'

Elisabeth eyed him with a cheeky grin.

Halley hid behind his tankard and fell silent.

Jan returned carrying a large vellum portfolio.

'Not talking?' he asked. 'I thought you'd jump all over him, Elisabeth – it's a change for you to have young company.'

Without waiting for an answer, he opened the book for inspection. The parchment was covered in neat calculations that Halley saw at once were astronomical in nature.

'These are Kepler's original manuscripts,' explained Jan. 'I purchased them. They are the only remaining memorial to him.'

'Only?'

'His grave is lost. Destroyed in war.'

Halley turned the pages. In places there were large, frustrated crossings-out and glossy stains of spilt wax. On other pages the writing became progressively sloping and lazy. *Where Kepler had been working into the night to finish*, thought Halley. He continued to turn the pages, marvelling at the sheer tenacity of the work.

When he lifted his head, his eyes looked straight into Elisabeth's gaze. She was watching him from behind her husband's back. As they locked eyes, she fluttered her eyelashes and looked away.

Once the night had settled itself over the land, the three of them headed for the observing platform. As they stepped on to the rooftop construction, Halley could not help but think back to Hooke's rickety perch. This one was as spacious as the other was cramped. It spanned three buildings.

'Don't your neighbours mind?' asked Halley.

Elisabeth suppressed a giggle.

'Neighbours?' boomed Jan. 'I *am* the neighbours. I own all three houses.'

The wooden planking creaked as they moved, audibly more under Elisabeth. The river that had borne Halley to Danzig was just visible between the buildings, rippling in the moonlight. Jan led them towards an enormous metal wedge that pointed straight up into the sky.

'I've never seen a sextant of that size,' said Halley.

'It's as Tycho Brahe would have used,' said Jan. He eased off the clamps and began to manoeuvre the metal contraption. To his chagrin, it squeaked. 'I curse the damp here. We fight to keep things working.'

'You model yourself on Tycho Brahe?' asked Halley.

'Of course. I consider him the greatest naked-eye astronomer who ever lived.' Jan turned to his wife. 'Let us show our young friend here how we work.'

They took their stations at different ends of the sextant and tipped the framework into position. Elisabeth bent to look through a small slot that guided her sight along the six-foot arm of the wedge to a similar slot at the other end. She turned a crank and the instrument moved a fraction.

Halley moved closer and indicated the eyepiece. 'May I?' He bent to the sights and pressed his eye to the metal. The star seemed naked without the glass of a lens between it and his eye. 'Is it always a two-man job?'

'Two-man, Edmond? Have you mistaken me for a man?' Elisabeth's breath was hot on his cheek. She smelled of flowers.

Halley looked round. 'No, I meant nothing by it.'

Jan's throaty laugh split the night. 'Stop teasing the boy, Elisabeth.'

Masked by the darkness, Jan could not possibly have seen the look that she directed at Halley. Even the young astronomer looked twice before he believed it himself. Desire glowed in her eyes. He jerked away as if stung.

Jan was oblivious, crouching beneath the sextant, peering along the moveable armature and adjusting it into position. His long coat pooled around him on the floor. 'Good . . . good . . . yes. Here,' he indicated his place at the instrument, 'all is aligned.'

Halley took his place; another anonymous star had been caught in the sights.

'Now we can read off the angle between the two stars,' Jan was saying, but Halley was finding it difficult to concentrate. He was completely distracted by Elisabeth's presence. 'We use mural quadrants now,' he said, as much to control himself as to inform his host. 'We mount them on walls built to face north–south. They have a single telescope on an armature, and we take the altitude of the star as it crosses the meridian line. We can do the rest with computation.'

'I read about your mural quadrants,' said Jan. 'One of Flamsteed's letters said that Hooke's design was dangerous. The armature so heavy it nearly had his student's fingers off. I'd prefer to lose a few arcseconds in accuracy than my own digits.' Hevelius waggled his fingers. 'I'm too old to change. I observe the way Tycho did, and perhaps that does make me a dying breed, but I'm as accurate as any newcomer. You have brought your own sextant, have you not?'

'I have,' nodded Halley, remembering the sailors' complaints as they had unloaded it that afternoon.

'We will have the servants bring it up tomorrow. We will observe side by side and compare results. You will see that I am as accurate as I say. Then you will take that message back to Mr Hooke and the doubters at the Royal Society.'

'I will, sir.'

'Jan, call me Jan.'

'I will, Jan.'

'Good. Then let's get some rest.' He curled a paternal arm around Halley as they left the roof. 'The good thing about being an old astronomer is I no longer need much sleep. But for five hours, nothing can wake me.'

An hour later, Halley was testing the softness of his mattress when the door creaked. *Elisabeth*. His breath caught as she slipped into the room and placed a soft finger over the question that had sprung to his lips.

Her eyes were big and round.

Halley's heart pounded.

She delicately removed her finger.

When he said nothing, she closed the door and turned to face him.

10
London

With a shudder of revulsion, Newton stepped over the litter and excrement that lined the street. He glanced down the dark alley. Yes, he was sure this was the one. He had committed the journey to memory in the dingy tavern room before dropping the map he had been sent on to the fire.

He hated London. Being here in the midst of such decay felt like crows pecking at his flesh. Even in the finest streets you would often glimpse a dead body. Mostly these were animal carcasses, but it was not uncommon to come across a human who had perished in the night. Yet, if he was going to exorcise the dreams that continued to haunt him, he could see no other way than to make this trip. He had made a breakthrough with his experiments, but he needed confirmation.

When he found the doorway, Newton squared his shoulders and rapped sharply.

A panel opened. There was a faint glow from inside and a silhouette. 'Yes?'

'I'm–'

'Don't tell me your name. Didn't you read the instructions? Show me the letter.'

Newton had planned to quote the pseudonym he had been using on his letters, but he was so startled by the admonishment that he said nothing. He retrieved the cryptic invitation from his pocket and pushed it through the panel. Every inch of him was poised to turn tail and run.

The door opened. Newton's blood froze. The doorman was hooded, like an executioner. 'Don't be shy. Come in.'

Newton stepped inside. The man closed the door, then touched a candle to the letter before dropping it on to a metal plate. They were in a small ante-room, little more than a closet. Pointing to a doorway, the man said, 'Down there.'

Newton peered into the gloom. 'There's no light.'

'There's only one way to go.'

Newton adjusted his satchel across his shoulders and descended. The door shut behind him. He was suspended in pitch darkness with only the touch of the cold wall on his fingers to reassure him that he was alive. He felt his way to the base of the staircase, where he discerned a door by the faint glow underneath it. He groped for the handle and pushed the door open. In the middle of the room, sitting on an earthen floor, was a single candle. The room smelled of damp earth and stale humans.

'Sit down, please,' offered an incongruously cultured voice.

There was a simple wooden chair and Newton perched on it. The candlelight revealed little. A metal grate behind it blocked most of the illumination from striking his hosts, though Newton could make out shapes sitting in a semi-circle.

As his eyes adjusted he saw shoes and legs: one pair was long and slim, another was in mud-splattered hose, yet another grossly swollen, with varicose veins bulging beneath the stockings. All wore the fashionable block heels. So, all were gentlemen.

And all are secret adepts. Newton's pulse raced.

The same voice cut through the murky air. 'You have been invited here today because it has come to our attention that you have made substantial progress in the alchemical arts. This is not an easy thing to undertake alone. Our art is much mistaken by those who do not truly understand what we seek. What is it that drives you?'

Newton cleared his throat. He was confident on this: alchemy was not the petty quest for gold but the pursuit of the lost knowledge of creation. 'I seek the bridge between the material and the spiritual worlds. Through my studies I pursue the active principles of nature, the ones that God has used to bind up His creation and make all transformations possible: why the clouds turn to rain; why life springs from dirty Earth. Transforming lead into gold? Yes, that would be a profound transformation, but not for the sake of wealth. It signifies the raising of baseness to perfection, as life signifies the perfection of inanimate matter.'

'Are you pure of heart in these intentions?'

'I am,' said Newton. 'I have brought something to demonstrate this to you.'

He opened the satchel and pulled out a small, ruby-coloured mass, which he placed in the palm of his hand. Stubby crystalline branches stuck out from the central blob in parody of fingers. There were gasps from the invisible committee.

'Where did you get that?' asked one.

'I made it in my own furnaces,' said Newton, basking in their surprise.

'But it's . . .'

'The star of regulus.'

A pair of liver-spotted hands reached out for the crystal. 'May I?'

Newton held back deliberately before handing it over. It allowed him to catch a glimpse of a lined face with sagging flesh.

Robert Boyle – as he had suspected.

'You did this without guidance?' asked Boyle.

'I used only my own reading into the subject and experiments.' He had decoded the recipe from an engraving of two winged dragons fighting. One was wounded and scorched and plummeting from the sky, and he had comprehended that he needed to pour the heated chemicals into a flask of liquid. So vivid had been his understanding of the allegory that he was starting to think of his experiments in terms of mythical beasts: griffins and feathered serpents and dragons – they were all coded references to alchemy.

There was an urgent round of muttering. They passed the crystal between them, then all the shadows stood. A new voice spoke up, younger and bolder.

'You are to be adopted forthwith. Stand, please.'

Newton did so. He was passed a Bible.

'Do you promise and swear faithfully and truly to keep close and conceal all secrets whatsoever shall be revealed to you by us, the Fathers of this Society, to which you are about to be admitted, so help you Almighty God?'

'I do,' said Newton.

'Do you promise and swear faithfully and truly not to reveal the names of any persons revealed to you, so help you Almighty God?'

'I do.'

'Then you are now one of us. What is the name you have chosen for yourself?'

'Jehovah Sanctus Unus.' Newton had constructed it from the Latinised version of his name. When he had seen the pseudonym hiding in the letters, he knew he had been sent a heavenly message. The realisation had chilled and thrilled him in equal measure.

One true God.

11

Rome, Papal States

Halley woke with images of Elisabeth Hevelius crowding his head. He could still see the hungry look in her eyes as vividly as if she were standing over him now, and he wondered why her image should return to him today. It had been months since he had left Danzig and begun his meandering journey across Europe to end up here on the Italian peninsula.

He was sticky with sweat, and rose to clean himself in the washbowl. The morning light was already overwhelming and carried a fearsome heat through the wooden shutters. Shaking water from his hair, he decided to breakfast in the piazza.

There was an old man watching the entrance of the English College building as Halley left that morning. His grey head was too large for the rest of the body, and he needed to support himself by resting on the fountain's wall.

Halley noticed him next to the sparkling water but thought nothing of it.

'Signor, a word with you, please?' It took a number of repetitions before Halley realised he was being addressed. The old man drew near with slow but certain footsteps. His features were carved as boldly as a sculpture, with a wide mouth and Roman nose set below a heavy brow.

'You are Edmond Halley, are you not?'

'I am,' said Halley warily.

The old man bowed. 'Your reputation as an astronomer precedes you.'

'Thank you, but you have me at a disadvantage.'

'My name is Vincenzo Viviani. However, I think that you would best recognise me if I tell you that I was Galileo's last student.'

Halley looked the man up and down, from the bandy legs to the disproportionately large head.

'I studied with the maestro in his final years.'

Halley found his voice. 'Have you had breakfast yet, Signor Viviani?'

They ended up eating pastries with coffee in a shady alcove on the perimeter of a packed market square. Oblivious to the commotion, Halley fixed on the lined face before him. 'But his final book, the one on motion, that he wrote after his trial, how did he arrange for its publication when he was being so closely watched?'

Viviani nodded as if it were an expected question. 'You have to understand that, by that point in his life, Galileo had nothing left to live for. He was taken in by Archbishop Piccolomini in Siena,who agreed to be his first gaoler. He nurtured Galileo back to health, but the maestro's rehabilitation there lasted just five months. Then the Vatican found out that the archbishop was a supporter, inviting academics and thinkers to revive my master's fractured mind, and they ordered Galileo to be moved. Piccolomini was ignored by Rome from then on, all influence lost. He remained in Siena as archbishop for the rest of his life. But he had given Galileo enough succour to continue. Without the archbishop's love, the maestro could not have endured the next burden.'

Halley nodded for Viviani to continue.

'He was placed under house arrest back in his home in Florence. Forbidden to see friends or host any gathering of more than a few people, he lived in peace, visiting his daughters in the convent nearby. But within weeks, his most beloved eldest daughter succumbed to dysentery and was taken by the Lord. On the day of her death, Galileo was escorted home by a local doctor, only to find an Inquisition officer waiting for him. He was told that he risked being taken back to Rome and imprisoned if he did not cease the calls for his good name to be restored.' The words were pouring from him now.

'So he wrote everything he could in his new book while he still had the time. And everything he wrote was demonstrably true. Anyone with the correct education could read and reproduce his experiments. No one could argue with him any more. His *Discorsi* is the greatest work of Italian philosophy ever written – why, it is a new philosophy! Yet even now, Italy does not know of its greatest son's achievement. His works are banned, his name not even whispered.' A heavy sigh escaped Viviani and his shoulders slumped. 'I vowed that I would see his grave moved here to Rome. It's unmarked, in a small room under the bell tower in Santa Croce. But I have failed. Galileo is still considered a traitor.'

'Not in northern Europe,' said Halley quickly, 'not in England.'

Viviani was breathing hard. 'I have covered my house in a tribute to his work. Etched into the stone on either side of the entrances are eulogies; above the door is a bust. Great men must be championed, no matter what the personal cost, don't you think, Signor Halley? They come along too rarely to be ignored. In 1564, the maestro Michelangelo died and Galileo was born. I believe they shared the same spirit. But who has it now?' Viviani looked around at the market stalls, the hordes of shoppers and hawkers. 'Who in this world now holds that spirit? No one in Italy, that's certain. But perhaps in England, with the Royal Society?' He fixed Halley with a burning stare. 'Tell me, Mr Halley, is there anyone in England who could be the next Galileo?'

Halley was still mulling the question over as he returned to the English College that evening.

'Letter for you, sir.' A steward waved the missive in the air.

Halley's eyes widened as he read.

'Is something wrong?' asked the steward.

It took a moment for Halley to form the words. 'I'm to return to England at once. I'm getting married.'

A black carriage was waiting at the bottom of the gangplank when Halley stepped on to the wharf at Greenwich. He made to step around the vehicle, but the scraping as its window was lowered drew his attention.

'Lift home?' The nasal voice was familiar.

Halley regarded the weasel-like features.

'Mr Winslow. How did you know I was arriving today?'

Winslow remained inscrutable.

Halley climbed into the carriage, which jolted into motion.

'So?' Winslow drawled.

'So what?'

'Anything to report? Don't tell me you've forgotten your promise to the King.'

'Of course not. Let me see . . . There were more than 24,000 recorded deaths in Paris last year, as opposed to 20,000 in London, but the same rate of marriages and births. So, to maintain the population each French couple must have four children.'

Winslow looked bored.

'I measured the Roman foot against our London one. It's four-tenths of an inch shorter. And the Greek foot is one-twentieth of an inch longer than we are used to.'

'Don't tell me that's it?'

'I think you will find that enough, especially if you're marching an army across Europe. One would want to know distances to the inch.'

Winslow scratched the stubble on his jaw. Then he struck the carriage roof with a brass-topped walking-stick and flicked his head towards the street. 'You can walk from here.'

On his wedding day, Halley found himself wondering if the unpleasant little man was watching from some dark nook during the ceremony. If he were spying, he would have seen Halley looking upwards to say the holy words, but would he have guessed that the astronomer was saying them to his mother rather than to God? He imagined her looking down and watching the proceedings with her soft smile.

Joane looked insufferably smug, as though she had single-handedly brought a miscreant to justice. While she stayed on the periphery of the celebrations, thankfully Halley's father took centre stage. Halley could not remember the last time he had seen the patriarch laugh so much, or drink so much for that matter.

Robert was there with Grace. She had regarded Halley so openly that at times he felt quite uncomfortable. Mary's sisters, Margaret and Dorothy, also attended; Margaret was the eldest, and she had gazed plaintively all day at Mary's gown. When the time came to be bundled up to bed, Halley's mind whirled with excitement and alcohol as Margaret and Dorothy flattened him on the soft mattress and stripped him of his stockings. Next to him, his friends were removing Mary's undergarments. Then the bawdy attendants all stood with their backs to the bed and flung the garments over their shoulders to see whose would reach the bride and groom.

All of them missed, but Halley caught Margaret's fabric talisman and dropped it quickly on his torso.

'There, see, you shall be married,' said Dorothy, to her elder sister's delight.

Then Halley and Mary were alone. They rolled to face each other . . .

12
Islington Village
1683

With Mary's eyes shut Halley could admire her without the usual self-consciousness that overcame him. Her fine blonde hair curled round her face and she wore a permanently mischievous expression. She had a little pointed chin, and pale lips that always seemed to be dancing on the edges of a smile.

Every time he studied her like this, it reminded him of their wedding night and the first time he had really looked at her, lying on the pillows beside him at the end of the day. The memory still shone so brightly that it was difficult to believe it had been all those months – a couple of years! – ago. He fancied it would never fade.

He backed out of their cottage, guiding her into the dusk by her slender forearms. The aroma of summer honey-suckle surrounded them. 'No peeping, Mrs Halley.'

She giggled as he led her away from the pool of candle-light at the back door and down the garden path.

'You'll spoil your surprise if you peep, and I'll have to punish you for being a bad wife,' he said with mock severity.

She immediately opened her eyes and pulled a comic face at him in challenge. Point made, she squeezed them shut again and continued with the game.

Halley fought the urge to kiss her and pull her back inside to their bedroom. By sheer willpower he led her to the bottom of the garden. *She had to see this – everyone should see it – to understand.* He guided her to a telescope pointing at the full, rich Moon. 'One more second,' he told his wife. He checked the telescope's alignment, making a whisker of an adjustment, and then guided her into the seat he had positioned. He crouched behind her and, cradling her jaw, moved her head to the telescope. 'Now then, open your right eye,' he whispered into her ear.

She gasped. 'Oh, it's beautiful! The hills are in a perfect circle. What a divine sight.' She cocked her head to check where the telescope was pointing. 'This isn't one of your tricks, is it? I'm really looking at the Moon?'

'It's real. It's called a crater. The Moon's covered in them, but we don't know how they got there.'

She returned her gaze to the eyepiece. 'Everything is silvery. Where are the plants and animals?'

'We don't know. It's a long way away – of course, not as far as the planets, and then the stars–'

'Stop it, Edmond, you're scaring me. I cannot imagine it.'

'No one can, but it's nothing to be scared of. It's a marvel.'

'It's truly the work of God . . .'

Halley said nothing.

'No wonder you spend so long out here.'

'I'm not just gazing around,' he said. 'We need to be able to fix our position at sea. You can measure your latitude easily by the height of the Pole Star above the horizon, but longitude is still difficult. The best way is to use the Moon. It takes twenty-eight days to orbit Earth, and as it does so, it passes in front of the stars that are scattered along its path. The exact time for each star to be eclipsed depends on your position on the Earth. What's funny?'

'You sound like a schoolmaster.'

'Then be a good pupil and pay attention. If we can compute a set of tables that gives the precise times when these stars will be eclipsed at Greenwich, then sailors can compare the time at which they observe the eclipses and calculate how far away they are from London. You see?

That's what I'm trying to do . . . but I need the Moon's position as precisely as I can measure it, night after night after–'

Mary twisted in the chair and threw her arms around him so tightly that when he stood up, he pulled her out of her chair and into his arms. The Moon forgotten, her slight frame was easy to carry back into the house.

In Hertfordshire, the soldiers arrived early to arrest the Earl of Essex. The rosy tints of morning bathed the redcoats as they marched through the estate's regimented gardens. They negotiated the box hedges and sidestepped the lavender, making for the original Tudor wing with its crumbling red bricks and diamond-shaped chimneys, before forcing their way into the slumbering mansion. On locating the Earl's sleeping-chamber they ordered his servants to dress and wig the peer. Then they carted him to the Tower – so the story went. It was all the coffeehouse patrons could talk about.

'When did it happen?' asked a wide-eyed merchant between hurried puffs on a clay pipe.

'Two days ago,' said the fat gossipmonger, basking in the attention.

'On what charge?'

'Conspiracy to assassinate the King and the Duke of York – they planned to strike at the Rye House Estate as the King was returning from Newmarket races, back in April, and they only failed because there was that fire at the racecourse, and the pair left early.'

'What did Essex hope to gain?

The fat man leaned forwards to draw the others closer. 'Parliament would crown the Duke of Monmouth.'

There was a collective muttering.

'But the King has never acknowledged him as a legitimate son.'

'Maybe not, but Monmouth is assuredly Protestant – unlike York – and by placing Monmouth on the throne, Parliament would be able to control him.'

'Monmouth agreed to this?'

The fat man placed a hand on top of his belly. 'The King has exiled him to Holland. Draw your own conclusions.'

The merchant turned, snatching the glowing pipe from his mouth. 'Mr Halley, your father works at the Tower, does he not?'

The astronomer nodded cautiously, reluctant to be drawn into the conversation, although he was as eager as the rest to hear what was being said. 'He's a Yeoman Warder in his spare time.'

'Then we can rely on you to inquire into this.'

Others turned now, undisguised eagerness in their eyes, and Halley wondered with some trepidation whether Winslow was lurking within earshot.

He put down his unfinished coffee. 'No gossip from me, gentlemen. *Au revoir, messieurs*, I have a night's observations to plan. Darkness is a shy visitor at this time of year; it's best to meet her prepared.'

Passing the city walls, Halley allowed a few seconds for the ripe smells to fall behind before unhooking the window-cord and letting the breeze into the carriage. The wheels slipped into the furrows of the country road, but before the rippling wheatfields gave way to the familiar cottages of Islington the vehicle came to a premature halt. Halley leaned out of the window and groaned. They were at the back of a queue of carts and carriages stretching into the village. Men were standing up in their carts trying to see what the hold-up was, while others walked in circles, impatiently waiting to be on their way.

Halley opened the door and jumped to the ground. 'I'll walk from here.' He thrust some coins at the driver, who tapped his temple. Setting off along the road, Halley shrugged off his jacket, flipped it over his shoulder and rolled up his shirtsleeves.

As he drew close to the cottages, with their gardens full of swaying hollyhocks, he spied his own top-heavy house, its upper floor balanced on the bay windows of the lower. In front of it, blocking the road, was his father's carriage. Its chipped door was hanging open. Mary dashed from the front door, holding her petticoats away from the dirt. 'Your father's here.'

'So I see.'

She leaned in close. 'He appears not to be himself.'

From the hallway emerged his father, plump and waddling, dressed in his blue Tower uniform.

'Father, you know you're bringing Islington to a standstill?'

'We must go.' Halley senior avoided his son's gaze.

'Go where?'

His father did not reply but turned to face the open carriage door.

'I have preparations to make for my observations tonight.' Halley searched his father's face for some inkling of a reason.

'Please,' his father hissed through clenched teeth, looking pointedly at the carriage.

Halley exchanged a glance with his wife. 'I'll be back presently . . .' He addressed his next words to his father. '. . . when this mania has passed.' He stalked to the carriage and settled on the lumpy upholstery.

His father sat opposite but said nothing until the noise of the wheels and the horses had risen. 'The Earl of Essex was murdered today,' he said, eyes downcast.

'You mean he was executed.'

'No, the trial didn't go ahead. There was no conviction, but the Duke of York had him murdered.'

'Have you finally fitted yourself for Bedlam? You're not even in your dress uniform. How could the Duke have visited?'

'We weren't told of the visit until we arrived this morning. The King wanted no pomp.'

Halley's voice grew sharp. 'The King, now.'

His father looked up, his dark eyes clear. 'Getting angry with me won't change anything. Essex was murdered by the Duke of York.'

'You witnessed this?'

'I didn't need to. Thomas Redman – you know, the young chap – was guarding the Earl. He saw everything. Everyone was panicking over the royal visit. I was sent to inspect the Duke's chambers and had called for fresh fruit. I was waiting for it to arrive when I heard the footsteps. It was the Duke. He burst in, courtiers with him, and he was livid, shouting about his brother being weak-minded. I couldn't move. They were talking about the King's intention to secure a conviction and then pardon Essex to disarm his supporters.' The old man wiped the sweat from his brow and continued. 'There was something about the evidence being circumstantial and everyone wanting to be cautious, what with the collapse of Titus Oates's case and his pack of lies about plots. But the Duke wasn't having any of it, especially as Essex had been arguing for his exclusion from the throne. Then the Duke became very calm. You see, he calculated that Essex had no idea he was to be pardoned. As far as the prisoner knew, he was going to be executed and forfeit his estate, leaving his family destitute. One of his courtiers, a horrible man, whining voice . . .'

A chill passed through Halley at the description of Winslow. He almost heard the man's drone as his father continued.

'. . . said they could tell the Earl that there was a noble way to ensure that his family retained their home. It was that a corpse couldn't be tried, so if Essex were to die before the trial, his crime would die with him and his estate would pass to his son. Next thing, the Duke reached into his coat and handed the man a pocket-knife. 'Take this to Essex,' he said. 'He might like to pare his nails before the trial begins.'

'That's when the Duke and his man noticed me. I didn't dare meet their eyes. I just waited until they'd gone. Thomas Redman told me the rest later. The Duke's man talked to the Earl in private and left. Soon after, they heard the Earl fall and found him in a pool of blood, his throat slit.' The old man's bloodshot eyes welled with tears. 'I'm a marked man with this knowledge.'

'Be calm, Father. There will be a solution.' *What am I saying? What solution?*

'This confirms every fear I've ever held about a Catholic monarch. England will be ruled by tyranny again,' said his father.

'We must put our trust in Parliament to curb him,' said Halley with deliberate calm.

'Tell me of the coffeehouses. Are there people willing to stand against the Duke – when the time comes?'

'Father! You forget how much we owe the House of Stuart. You must talk to no one about today's events. Put this far from your mind. Let's turn the carriage around. Come and dine with us tonight.'

The old man's head sank into his hands.

'Tell me you have not already opened your mouth.'

His father spoke in a whisper, speaking through the cage of his fingers. 'It was all the Yeomanry could talk about. Mr Redman was eager to give his account, and upon his mentioning of the Duke's man, the others turned to me. I told them I heard nothing, but they kept asking. Question after question after question.' He looked up entreatingly at his son. 'They are men of discretion.'

'They are men of discretion with wives.' Halley grasped his father by the shoulders. 'Forget everything you have seen and heard today, Father, for all our sakes. From this moment on, we must never speak of this again.'

13
London

Grace twirled. 'How do I look?' Her dress was silver-grey in colour, with three-quarter-length sleeves that displayed her slender forearms and wrists, and a pleated skirt with plenty of fabric at the back. She flicked open her fan and hid behind it, peeping out over the top.

'You look beautiful. But you know I've always thought that. Now, come, let's be on our way.'

'Wait,' she called. 'We cannot go yet.'

'Why ever not?'

'You still have the muslin in your hair.'

Hooke reached up to the curling strips she had tied in after cutting his hair that afternoon. With a curse he began to pull at them.

'Here, I'll do it,' she said.

Once on their way, a whirl of motion in the corner of Hooke's eye caught his attention. Behind an open gateway, a small gang of apprentices was gathered in a wood-yard. The one in the middle was swinging a bucket of water in a complete circle, astounding the others with the way the water seemed to be glued to the inside.

Hooke approached the gateway and the boys stopped their play.

'No, do it again. Please,' Hooke called to them, ignoring the elder boys who were gawping at Grace.

The boy with the bucket shrugged and set it back into motion. Hooke watched the way the boy flicked his wrist and the way the string went taut to drag the pail into a circular motion.

Fascinating. Like an orbit.

'Oi!' From a side door, a stout man in a filthy work apron swaggered into the yard. 'I don't pay you to do that.'

The bucket's string sagged and the distracted boy took a full drenching. The other apprentices scrambled, torn between the hilarity of their friend's soaking and their own desire to avoid a punishment.

The yard's owner noticed Hooke standing in the gateway and turned to face him.

'Come on, Uncle.' Grace pulled on his arm.

The milling crowd ahead meant they were nearly at their destination.

'Walk like a statesman,' said Grace out of the side of her mouth. 'You built the place; act as though you owned it.'

Before them, the ornate exterior of the Dorset Gardens Theatre was shining in the evening sun. With its colonnaded entrance and life-sized statues on the second-storey balustrades, it was exactly as he had envisaged. He had drawn the plans in candlelight when all was quiet in the city and anything seemed possible. But, of course, most people assumed it was another of Wren's achievements.

Heads turned as Hooke led Grace towards the grand entrance. He caught a number of the crowd nodding in their direction and passing comment with their neighbours, a few gentlemen attempting to catch Grace's eye. She acknowledged none of them, Hooke noticed with pride.

Standing to attention among the crowd, a large number of soldiers watched the proceedings hawkishly.

'Why are they here?' Grace asked.

Hooke grinned, ready to spring his surprise. 'The King is attending tonight. You will finally meet him.'

She straightened herself even more, and Hooke wished he could do the same. He guided her around the theatre to where the shallow Thames lapped the wooden quay of the river entrance, and squeezed into a gap between the other well-dressed city folk.

The King's boat was already in sight. It was a long vessel of honey-coloured wood, with a canopy to the rear, under which the shadow of the King could just be glimpsed. Eight oarsmen cut the river, breaking the surface into golden glitter, and behind them an entourage of smaller vessels bobbed in the wake.

As the King's boat neared the theatre, a uniformed boy leapt to the bank carrying a rope, and those on the quayside helped him draw the vessel to a dignified stop.

Dressed in a profusion of white, with a full wig of coal-black curls, the King stumbled once before stepping up on to the quay. As Charles Rex began to climb the small flight of stone steps, the onlookers bowed and curtsied.

The white-hosed ankles stopped just before reaching Hooke. 'Well, well,' said the King in his customary measured timbre, 'who have we here?'

'I'm Grace Hooke, Your Majesty. Niece to my loving uncle.'

'So you're Grace, niece to the man to whom bowing is second nature.'

Hooke looked up, forcing himself to smile. At this proximity, Hooke noted that it was obvious the King's eyebrows were loaded with kohl to mask the grey.

'So, this is whom you petitioned me about,' said the King, running his eyes over Grace. 'You sly old fox.'

'She is my closest family. Like a daughter to me, Your Majesty.'

'Is that right?' said the King in a loaded tone. He paused a moment, then said, 'Share the royal box with me tonight.' With that he swept off to the theatre, leaving Grace's mouth gaping and Hooke feeling uneasy.

'This way, sir, madam,' said an aide.

Once inside, the King insisted that Grace sit between him and Hooke. 'I must admit that I'm surprised to see you here tonight, Mr Hooke. I had not credited you with much of a sense of humour.'

'I hear that Mr Shadwell's play is a keen satire against those amateurs who pretend to know science.'

The King raised his painted eyebrows. 'That's one way of looking at it.' He then launched into animated conversation with Grace, smiling to reveal his remaining mustard-coloured teeth every time she stifled a giggle. Hooke could tell Grace was only acting the part – as she should with the King – but even so he grew restless. There were a number of nods and glances coming in his direction from the stalls, too. He had at first thought they were simply watching the King or Grace, but no, they were definitely aimed at him. While he was a well-known figure in the coffeehouses and across the city, this level of recognition was unusual.

A pair of stagehands carried on some pretend display cases. Hooke couldn't help thinking that the shoddy construction would not pass muster at the Royal Society. Then a young man in a flowing gown delivered a prelude consisting of an over-enthusiastic exposition on Lucretius and his fusion of philosophy and verse. Throughout its mercifully brief tedium Hooke found himself more concerned with whether the poor fit of the actor's gown was deliberate or not.

He forced himself to pay attention as the first act of *The Virtuoso* began and the audience was introduced to Sir Nicholas Gimcrack, a natural philosopher of supposedly great repute. Gimcrack was played by an actor who had clearly stuffed a pillow up his jacket to feign a paunch and daubed himself with soot to increase the lines on his too-youthful face. He had a habit of dragging a foot across the stage or hunching himself and rubbing his hands together, as if this were somehow a characteristic of age.

There was a scattering of laughter as the characters discussed a book of lunar geology that Gimcrack had compiled, capturing the smallest lunar detail, during which his wife's numerous adulteries went on behind his back.

Hooke muttered to Grace. 'This is no substance for ridicule. Galileo's discovery of mountains on the Moon was a turning point in–'

A glance from the King silenced him.

Then Hooke's world collapsed.

Gimcrack began extolling the virtues of not one but two nieces, who duly filed on to the stage. One was a blonde, the other a redhead, each as lithesome as Grace. The theatre began to warp around him as the redhead announced that her uncle had spent all of his time and money on microscopes to study the tiniest of living creatures.

They were lampooning Hooke's own book, *Micrographia*.

'Yes,' agreed the blonde and with perfect comic timing she explained that Uncle had broken his brains over the nature of maggots yet never cared to understand mankind.

Hooke observed the audience in disbelief; they were roaring with laughter.

But we are recovering the lost wisdom of mankind.

Rosy faces were turned towards him from every direction. People were nudging each other and pointing. The sterner his expression, the more derision he seemed to invite.

Nearby laughter, deep and throaty, caught his ear. Gripping the arms of his chair, Hooke turned in its direction. Beyond Grace's downcast eyes, Charles was guffawing. The King glanced over and lifted those disgusting eyebrows. 'Congratulations, Mr Hooke, you have made it into the upper echelons of London life.'

'That's right! You can all point and laugh.' Hooke pushed through the boisterous crowds outside the theatre at the play's conclusion.

'Come along, Uncle.' Grace propelled him gently in the right direction.

'I have never been so embarrassed in my life.' Hooke glared from one patron to another.

'They're only jealous. Let's not give them any more cause for amusement.'

Once they were away from the crowds, Grace slipped her hand into his. It felt warm and the heat spread like a balm through him.

'Did you really petition the King on my behalf?' she asked.

Hooke nodded. 'Your father was in debt to the Newport Commission. The Crown was within its rights to confiscate your home to reclaim the money. What else could I do but ask the King if I could manage John's estate? Otherwise, you and your mother would have been homeless. But I didn't know at that stage about your . . . condition.' He glanced at her abdomen, hidden beneath the fine embroidery of her dress. 'I went to court to plead for the estate, only to discover your lover had beaten me to the same proposition.' He failed to keep the edge completely out of his voice. 'That was how I learned you were with child. The King told me everything, as Sir Robert Holmes had told him. It seems that Sir Robert was not as discreet a gentleman as he might have been. By the time I arrived, the King had already granted him your father's estate.'

'Does all London society know of my humiliation?'

He swallowed the rebuke that jumped to his lips.

'Don't be cross with me, Uncle. Sir Robert was the only person I could turn to. He took the child as soon as she was born. I never even saw her, not properly. Just heard her cry as they carried her away. He made no secret that Mary was his – that's what he called her, Mary, but the name was my suggestion . . .'

Hooke squeezed her hand. Something was taking hold of his senses. The world seemed to be twisting around him.

Blinking did not help clear his vision. It was similar to the confusion he had suffered after being attacked near the Letter Office, but this time there had been no impact. Nevertheless his mind had deserted him. He could hear Grace's voice but not make out her words. He could tell that she was agitated but could do nothing to reassure her; his voice had left him. He staggered into something hard – a wall? It was difficult to tell which motions were real and which imaginary. Whatever had gripped his mind had taken complete control.

Now he could hear Grace shouting for help, but he was utterly unable to respond. He concentrated on the solid mass of wall pressing against his back. Having lost all sense of position, he could not be certain whether he was horizontal or vertical.

The night was a tempest of colours, the kind of mess that he had seen Grace make on a painting palette once when she had lost her temper. He watched them swirl around him until gradually things cleared. He was upright, leaning against the wall. Grace was approaching at a run with a gentleman in tow.

Hooke raised his hand to halt them. 'I feel quite well again, thank you.' A sense of elation and a strange clarity of thought entered his mind.

'You're quite well, sir?' asked the gentleman.

Hooke nodded. Puzzled, Grace thanked her helper and made to put an arm around his shoulders.

'There will be no need for that,' he said mildly, tucking her arm through his.

'You scared me. Are you sure you're quite recovered?'

Hooke nodded. 'Whatever it was has passed.'

They set off.

'I want to thank you for taking me out tonight.' A small laugh escaped her. 'Why did you never marry? You'd have made someone a good husband, you know.'

'It's against the terms of my employment at Gresham. Besides, wives are a distraction.'

'I think I'd have liked marriage, and a family.'

'Galileo's daughter Virginia never married. Instead, they remained constant companions.'

She dipped her head to look at him through her eyelashes. 'Is that my role in this: daughter to your Galileo?'

'Hardly. She became a nun.'

Grace giggled, and the sound of it charged him with longing.

'I love you,' he said, suddenly serious.

She pulled him a little closer and they walked on.

14

Winter fell upon the city. The first bite sent carriages and people slipping across icy puddles. The second blanketed the streets in snow, stifling the sounds and robbing them of smells. The initial covering was pristine, but as a hundred thousand home fires began burning in grates across the capital, so the smoke mingled with the clouds to create a sulphurous cloud, staining the snow to bile-yellow. The heavy cloaks and jackets that people had taken to wearing transformed the population into black ghosts, floating through the muffled city.

The Thames froze gradually, cracking and splitting at the margins until it came to a vitreous halt. Incoming carpenters and stallholders went to work building counters and booths to transform the stilled river into London's latest, albeit temporary, thoroughfare. The wherries were fitted with skis so that the boatmen could drag their passengers across the ice instead of rowing them through the water. The crowds loved it, picking their way with care around the bull-baiting and the young boys who slid along the frozen surface, all the while trying to ignore the numbing cold that crept upwards.

On this particular evening, Halley was on his way to Garraway's coffeehouse. He hurried through the tainted snowfall, long cloak rippling. The excitement of the Frost Fair had been eclipsed by the frustrations he had felt with his mathematics today. He had spent the day huddled in his study, shivering against the air that rolled off the window. He had chased numbers across the page, formulating and reformulating but never making any progress. Now he had to exorcise the frustration before it gave him another sleepless night.

Garraway's coffeehouse was a long hall, full of aromatic warmth and the hubbub of conversation. The well-heeled squeezed on to wooden benches alongside those on the make. Some chatted, always eager for a new piece of information or the chance to turn a profit; others read the broadsheets that littered the place, though they contained little more than propaganda of one sort or another.

Halley placed a penny on the bar, where a woman with rouged lips and a gigantic wig stood among a castle of pots. The size of her wig was matched only by her impressive cleavage. She did not look round, but set down her cloth and indifferently handed over a shallow bowl at the sound of the coin.

'Thank you, Rose,' acknowledged Halley in his baritone.

Her head turned immediately. 'Well, Edmond Halley. Haven't seen you round here for a while.'

'I'm a married man now.'

She pulled a face. 'What can I do for you tonight?'

A smile tugged at his lips as he remembered the old days. 'What indeed?' He leaned towards her, lowering his voice. 'Alas, tonight I am here on a matter of some urgency. I'm looking for Robert Hooke.'

She pointed to the far corner. 'He's over there.'

'Thank you, Rose.'

'Perhaps later?' she called, raising an eyebrow.

Halley winked, instantly regretting the gesture. *That was how you used to behave*, he scolded himself. He stopped to fill his bowl from the black cauldron suspended over the fire, then made for the large gathering near the far corner of the room. Sure enough, Hooke sat at its focus.

Hunched in his seat, his leathery skin illuminated by the ruddy glow of the fire, Hooke was a shaman. His yellowing eyes held a group of men spellbound. Some stared, others nodded, and shoulder to shoulder they hung on his words as if they were an incantation. Tonight, the shaman's tale was of orbits.

'Kepler showed us that there is no mystery to the movement of the planets. Each heavenly orb follows an elliptical path around the Sun, each path as individual as the ugly faces I am looking at tonight.' Hooke swept his eyes around the crowd, relishing the smattering of laughter.

'Watch!' He overturned one of the broadsheets and dropped on to it a circle of twine from his hand. He fitted the parted arms of a metal calliper into the twine and, with the other hand, hooked an inked quill into it. Pulling the string taut with the pen, he traced a curve, allowing the tension of the string to guide his hand into an ellipse. He removed the apparatus from the paper and pressed the quill into the point left by one of the calliper's arms. 'The Sun sits here, at a focus around which every planet follows an elliptical path of some description.'

'But why an ellipse?' called one of the crowd.

Hooke pointed at the man as if pressing a button. 'That, sir, is the question of the age. Why the ellipse? Kepler found another clue: that a planet moves faster through its orbit when it is closer to the Sun. So whatever power the Sun exerts to move the planets must diminish with distance. But how does this force communicate itself across space?' Hooke pointed at the man again. 'Say I wanted to move you, sir. How would I accomplish that feat if I am fixed to this spot and cannot approach you?'

'I would say it's impossible, so long as I stay out of your reach.'

Hooke gave the man next to him a forceful shove, spilling the man's coffee and propelling him into his neighbour, who fell sideways into his neighbour. In a second, the momentum had reached the questioner, who rocked sideways.

'Mechanics, gentlemen,' said Hooke in triumph. 'It's the key to understanding how forces transmit themselves. They must travel through an intervening medium. For gravity, the medium is a substance not found on Earth: the ether.'

A member of the crowd called out, 'But what of the comets, Mr Hooke? Do they truly bring evil?'

'They are most certainly not evil. They're natural objects, not omens.'

'But one did appear over the city before the plague,' said another of the audience.

Hooke sighed. 'How then shall we measure their wickedness? There is no uniform lag of time between their appearance and calamity. And why did we not see a comet before the Great Fire, was that not terrible enough to warrant an omen? The world is a mechanical place. If the comets truly brought evil, they would do so without exception, just as a wheel always turns when pushed.'

Halley could contain himself no longer. He called out in a disguised voice, 'I've read that comets formed in the Earth's own airy heights, condensed from the ascending vapours of human sin and wickedness. And that this poisonous stuff then rains back on to us, forming all unpleasant phenomena, such as diseases, sudden death, bad weather . . . Frenchmen.'

Hooke swung his head in search of the heckler.

Halley nonchalantly curled a strand of his long dark hair behind an ear.

The two men eyed each other.

'You!' spat Hooke.

'Have you still not forgiven me for that business with Hevelius?'

'You sided with a foreigner against me.'

'I worked with him, used his quadrant. I know his methods. You just dismissed him out of hand.'

'It's not the only thing of his you handled, from what I hear.'

Halley looked to the ceiling. Around him, Hooke's impromptu audience began to drift away. When he looked down again, Hooke was gathering his things.

'I have to leave,' said the Gresham Professor.

'Not yet, your rendezvous with Sir Christopher isn't until nine o'clock.'

Hooke's gaze sharpened.

'I stopped by your rooms,' explained Halley. 'Grace informed me that you're to meet with Sir Christopher in Jonathan's coffeehouse at nine o'clock.'

Hooke stuffed the loop of twine into his pocket and rose from the table.

'Let me accompany you,' said Halley.

'You're wasting your time.' Hooke pushed through the crowded room and disappeared outside.

Halley sighed, then hurried after him. Just ahead of him, Hooke moved in and out of sight in the thronged alleyways.

'Robert, I must speak to you. Only you can help me.'

Hooke did not stop, but there was interest in his voice. 'How so?'

'I wish to talk of gravitation.'

Hooke paused before the coffeehouse entrance and turned to Halley. 'What about it?'

'Let's talk inside, it's a bitter night.'

'Pay my entry and you buy yourself some time.'

Halley reached for the door. 'After you, sir.'

'Gentlemen, I have been wrestling with a simple but vexing question,' said Halley once they had squeezed around the table that Wren had been guarding. 'Night after night I study the Moon. But what keeps it suspended in its orbit? We all agree that something called gravity holds us to the Earth, and makes things fall to the floor. So, if this gravity extends into space, what prevents the Moon from falling on our heads?'

Wren tapped the table with his finger, as if bringing a meeting to order. 'Gravity must decrease with height. So by the height of the Moon, its grip is so light . . .'

Halley stopped him with a shake of his head. 'Even if the grip is light, it will still attract the Moon. The closer the Moon gets, the more it will be pulled and the faster it will fall until . . .' Halley clapped his hands together.

'Then a balance of two forces, one pulling the Moon down, the other pulling it out into space.'

'Aristotle's opposing forces of gravity and levity? That is some old thinking, to be sure,' said Halley.

'Are we so modern that we now reject all old ideas out of hand?' asked Wren.

'Indeed we do,' said Halley, 'unless we can prove them. Besides, that would make the Moon hover rather than orbit. There must be something else at work that makes the Moon travel around the Earth but not fall to it. For the planets orbiting the Sun, too . . .'

Hooke was smiling to himself.

'Robert?' asked Wren. 'Do you have something to say?'

At length, Hooke spoke. 'I already have the solution. Gravity diminishes with distance in the following way: double the distance of two objects and the gravitational force between them quarters. Triple the distance and gravity falls by one-ninth.'

'An inverse square law,' said Halley.

'Exactly.'

'But that doesn't solve the problem of what keeps the Moon in orbit.'

'There's another force – but not pulling outwards. It tries to push the Moon along at right angles to the downward pull of gravity. Together, these two forces drive the Moon into its perpetual orbit around the Earth. It's similar to the way water stays in a bucket if the vessel is swung sufficiently fast above one's head. Just as the water defies the pull of gravity because of its motion, so too does the Moon.'

Halley searched Hooke's face for some hint of jest. 'How do you know this? Did Newton reply to your letters?'

Hooke looked insulted. 'No, he did not.'

'Of us all, he has the mathematical acumen to perform the analysis,' said Wren. 'From what I hear of his fluxions-and-fluents technique, it would be the perfect mathematical tool to use.'

Hooke stabbed him with a look. 'How many times must I tell people? Newton has given up on philosophy.'

'What happened to that little telescope he invented, the one that used mirrors instead of lenses? Didn't he present it to the Society at the same time he came to talk about colours?'

'It was impractical,' grumbled Hooke. 'The mirror tarnished every few months and needed polishing. Now he keeps his mathematics secret. He's of no help to us. There's no need for him to start meddling in any of this. Besides, Leibniz is beginning to publish his version of the fluxions, so we don't need Newton.'

'The Society needs as many new members as it can get. As it stands, we're ailing,' said Wren.

'We're a little slow on progress, that's all,' said Hooke.

'There are those close to Charles Rex who openly wonder why he continues to support us. But, if we could discover something – something useful – we might silence our critics,' said Wren. 'Or perhaps our little experiment with natural philosophy is over – perhaps we have plumbed it for all it has to offer.'

'Never! The world is there for us to measure. We have barely begun,' said Hooke. 'But we don't need Newton; I have pressed on alone.'

'Can you prove that elliptical orbits follow from an inverse square law of gravity?' asked Halley.

'Yes. Kepler's laws proceed from three suppositions.' Hooke counted them off on his crooked fingers. 'One, all celestial bodies have an innate ability to attract others. This gravity also prevents such bodies as the Earth from falling to pieces. Two, a body in motion will continue travelling in a straight line until some other power deflects it, whereupon it will describe a circle or an ellipse, or simply a part of a curve before it moves out of range. And three, the attractive power of an object diminishes with distance. The further away something is, the less powerfully it is affected. The decrease – as I have said – is in the inverse proportion to the square of the distance.'

'But can you back this up with a solid chain of mathematics, as you told me that Bonfire Night you had asked Newton to provide?'

Hooke looked into his steaming drink. 'Of course.'

'Then show us,' said Wren.

'Yes, show us,' said Halley. 'The solution would take us towards the perfection of astronomy.'

Hooke continued to contemplate his drink.

Halley exchanged a glance with Wren, who rolled his eyes.

'I will give you forty shillings if you show us the proof,' said Wren.

Hooke blew out a derisive breath. 'Hardly a prize worthy of winning.'

'Then think of the prestige,' said Wren.

'There are too few capable of understanding its importance, and even fewer who could make use of such knowledge,' said Hooke.

'Maybe so, but what do you gain from withholding it?' asked Wren.

Hooke looked up. 'Respect. When others have tried and failed, then they will appreciate its true value.'

'Then let us set a deadline.' There was impatience in Wren's voice. 'Say, two months from today. If no one has claimed the forty shillings by proving that elliptical orbits follow from an inverse square law of gravity, then you will enlighten us all.'

When Hooke failed to comment, Wren leaned towards Halley and spoke in mockingly conspiratorial tones. 'If still the solution is not forthcoming, one of us will have to visit Cambridge and plead with Newton to save us.'

Hooke grimaced at Halley's hearty laugh.

Later, Halley gathered his cloak and rose from the table. 'Gentlemen, I thank you for your time but I must start for home.'

'Will you not stay? The hour is only ten. Where is your passion?' Hooke called after him.

'There is passion aplenty in me, which is why I must bid you gentlemen good-night and find me a carriage back to Islington and my wife.'

As he neared the door a fresh-faced man with a chin sprouting inch-long blond whiskers blocked his way. The young man's eyes were unusually large, as if startled, and of the palest blue. He spoke, stumbling over his words. 'I would speak with you on a delicate but important matter.'

'Who are you?' asked Halley.

'Hugh Speke, a lawyer at Lincoln's Inn Fields.'

'Lawyer or student?'

'In my final year of study.' Speke's voice quickened but remained quiet. 'I wish to speak with your father.'

Halley said nothing.

Speke whispered, 'It concerns the Earl of Essex.'

'My father has no link to the Earl of Essex.'

'He was at the Tower the day the Earl was murdered. He attended the Duke of York.'

Halley pushed past the young man. Just before stepping into the welcoming night air, he risked a glance over his shoulder.

Speke was still watching.

15

Next morning, the sprawl of the tannery and the tight knot of dyers' buildings told Halley that his carriage was nearing its destination. At the front gates of his father's soapworks he took a deep breath against the cold and climbed out. On the horizon the Sun looked pale, frozen into the powder-blue sky. At least the frigid air disguised the reek of the surrounding factories.The gatekeeper greeted Halley and instructed him to head for the boiling-hall. Everyone wanted to keep warm in there today.

Halley stretched his arms wide. 'And who can blame them?' He felt not a shred of the jocularity he mimed. He had paced all night wrestling with what to do. He felt that he was betraying Mary by not confiding in her about the encounter with Speke, but was convinced that it should remain a secret for now.

'Oh, I nearly forgot,' called the gatekeeper. 'Your father has some gentlemen with him.'

Once inside the hall Halley spotted them immediately: two well-dressed gentleman strutting along behind his father on the gantry. One of the visitors was inspecting everything while holding a handkerchief to his nose; the other was bustling about with a book and quill, making notes and squeezing past the beet-faced workers who were stirring giant paddles through the stinking concoctions in the wooden vats below.

Each vessel was positioned on a flat iron plate. A heavy-set man was orchestrating a team to keep the trench underneath the vats fuelled with logs and blazing. Halley indicated the visitors with a questioning look. The fire-master shrugged and carried on directing his men, shouting to be heard over the crackling flames.

The trio made their way down the steps to floor level. Halley's father did a double-take at the sight of his son and flicked his thumb towards the offices at the side of the factory, little more than a row of wooden cubicles lined against the interior wall. Crude rectangular openings led into each of the spaces. Halley entered his father's office and leaned against the desk.

'Production is up, and we have new products specially for washing hair, and for gentlemen to use, too,' he heard his father saying as the three men passed by.

'For gentlemen? Will that really catch on?'

His wait stretched out and soon he found himself pacing. When his father appeared, he was alone and looked tired.

'What's going on?'

'Son, what a pleasant surprise to see you. I was thinking about you yesterday. Do you know it's been two years since your wedding? Is everything all right with Mary?' He bustled around the office, looking busy but really just moving papers around.

'Perfectly, why should she not be?'

'You know . . .'

'No, I don't. Mary and I are as happy as we could be.'

'Does the house not seem empty with just the two of you in all those rooms?'

'We have the housekeeper and my boot-boy to keep us company . . . Why, Father, I do believe you're hankering for grandchildren.'

'When you're away on your travels, they would give Mary something to do.'

'I plan no further trips. My lunar observations will consume me for years.'

'That's your problem. You spend nights out at your telescope instead of with your wife.'

'Only the clear nights.'

The old man looked blank, his eyes unfocused. 'Remember how you ran out into the garden with your mother's jewels when the Great Fire began, so I could bury them?'

It took Halley a moment to adjust to the new direction. 'I'd never seen you engaged in manual labour before – or since. You insisted we bury a case of wine, too. I could barely lift it.'

'It was expensive stuff. You cried when you found out that your school had been destroyed.'

'I didn't realise at the time that a better London would grow up as a result of the fire.'

'A grander London, for certain, but I'm not so sure about better.' His father sagged into the office chair. He spoke in a low whisper. 'Two men came here a few days ago, asking questions.'

'Speke? Blond whiskers? Blue eyes?'

His father nodded. 'And an older man named Braddon.'

'Did you tell them anything?'

'I didn't have to. They knew it all, even the pocket-knife. All they wanted me to do was confirm it.'

'I ran into Speke last night.'

'It's time for me to leave London. I'm too old for these intrigues.'

'You cannot run. It'll make you look guilty.'

'What could be more natural than an old man retiring?'

'Even so, it could be misconstrued,' Halley reasoned.

'It's too late. Those gentlemen in the boiling-hall are buying the factory, and there's a perfect property in Kent. I have a meeting set tomorrow afternoon with a broker. I haven't spoken to your mother . . .'

Stepmother, thought Halley.

'I know she'll be reluctant to leave London, but I plan to tell her this afternoon.'

'Rather you than me.'

Halley's father frowned. 'I know you haven't been comfortable with her, son. And I'm sorry about that, but . . .'

'Please reconsider. There are other ways . . .' Halley stopped at the sight of his father's upraised palm. It was trembling more than usual.

'Son, I've never understood what you see in the stars. I look up and see nothing but a pointless scattering of lights. You see reason, order, even meaning – enough meaning that I suspect you need nothing else to guide you. Now the tables are reversed. What appears chaotic to you is clear to me. Accept my actions.'

The fatigue from his sleepless night chose that moment to engulf Halley. He looked into his father's pleading eyes and nodded. 'Very well, Father, how can I help?'

Hooke struggled under the weight of an iron frame, wishing that the dawn were more advanced so that at least he could see where he was going. He squeezed himself and his burden through the doorway and with a grunt hefted the heavy cradle on to the dining-table. Pausing to regain his breath, he returned to the cellar with a candle and heaved out the wooden cone, setting off a small avalanche of boxes that he would have to clear up later.

The shallow cone was as big as his arms' span and he half-pushed, half-carried it up the stairs. With more heavy breaths he lodged it, point downwards, into the circular iron framework.

Already on the table was the trio of iron balls, his scattered diaries and a sheaf of old letters tied in a dusty ribbon. He thumbed through one of the diaries in search of the relevant passages from when he had used this apparatus before, growing more impatient with each page he turned. Finally he saw it, scrawled in the margin next to his description of the self-same experiment: *A perfect system of the heavens*.

His scribbled phrase brought it all flooding back: the growl of the iron balls rolling around the wooden cone, their beautiful elliptical trajectories and his lightning-bolt of recognition that their looping paths mirrored the way the planets moved.

That same evening, nearly four years ago now, he remembered hauling the apparatus down to Garraway's and repeating the demonstration to the delight of the patrons. No one there had disputed his conclusions; they just followed his reasoning and accepted his arguments. Then, later, he had seen the apprentice whirling the bucket and he had been as sure as he could be that it all fitted.

How could he transform such incandescent inspiration into sterile numbers? He threw the diary to the floor. It snagged the letters in its fall and burst them from their ribbon. Landing on its spine, the diary tumbled under the table and the letters fluttered down after it.

'What's wrong, Uncle?' Grace was peering at him with those enormous brown eyes.

'I'm fine,' he grumbled to cover his awkwardness.

'Let me help you.' She scooped up the scattered papers. 'Shall I put these back on the table?'

'They would be better on the fire to save their author from further embarrassment. Look at that.' He pointed to the topmost sheet. In the midst of the hurried, sloping handwriting was a hand-drawn circle. From this was drawn a simple tower, with a downward stroke of ink from its top that met the surface just in front of the tower, then passed through the surface to spiral into the central point.

'Wrong,' announced Hooke.

'Who drew it?'

'Isaac Newton – why they want him back I'll never know.'

'I don't even know what it's supposed to be.'

'The path of an object falling through space.' He scooped up one of the iron balls from the table and set it rolling around the upturned cone. 'That's the shape it should be.'

'Oh . . .'

'The perfect system of the heavens.' He pointed at the cone, 'Without the friction of the surface, these balls would loop back up to the rim in perfect ellipses – just as the planets do in their orbits. But Newton couldn't see that until I pointed it out to him.'

She smoothed the hair away from his face. 'Can I get you anything?'

'No, I've got to work. Some algebra – not my strongest suit, I admit, but unavoidable it seems in this brave new age.'

Later, when the mathematics had defeated Hooke for the day, a notion struck him. He gathered together some apparatus and hurried out of the college, leaving a bewildered Grace in his wake.

Heading for the tall white column of the Great Fire Monument, with its flaming gilt urn shining in the feeble winter sunlight, he wondered why he had not thought of this before.

He pushed at the oak door in its base and stepped inside. A spiral staircase snaked around the interior to a spot of light three hundred and eleven steps above. He inhaled deeply and began to climb, hugging the boxes ever more tightly. By the time he reached the cramped eyrie his knees were aching.

He placed his burdens on the floor, two wooden boxes and a length of string, and peered down the bore of the staircase. Two hundred feet below, in the darkness, was the ground; he was now at the highest point any man could be over London, double the height of the rafters in old Saint Paul's.

He secured a brass hook into a wooden beam over the drop, then bent to the larger wooden box. Nestled in the baize was an apothecary's scale balance. Hooke reverently lifted the instrument and hung it from the ceiling hook.

In the other box were the weights. He loaded a sixteen-ounce weight into each pan and balanced the string across the right-hand pan as well, smiling as the instrument tilted under the weight. He added more weights to the left-hand pan until they were level again.

He unhooked the right-hand pan, tied the string between it and the balance, and let it down the shaft, almost to the ground. The contents of the right-hand pan must now weigh more because they were closer to the ground, where the pull of gravity must be stronger.

He waited for the balance to settle, so he could measure the difference. It would take some minutes, he knew. He waited, not daring to look. Finally, he turned to the balance, blinking in disbelief.

It was perfectly level.

He touched it to make sure it had not stuck and waited again for the oscillations to die away. As before, it returned to perfectly level. The conclusion was inescapable: even two hundred feet above the Earth's surface, the decrease in the strength of gravity was imperceptible. He could gain no clue to help him this way.

He released the string and let the pan clang to the floor. He packed away the rest of the apparatus and began the downward journey. Somehow the spiral descent felt more wearisome than the climb.

Halley slithered along the watery street to his father's house just as the crescent Moon was beginning to set in the darkening sky. Around him, the dripping ice and snow seemed to hold on to the pallid glow of the moonbeams. He pushed open the front door and stepped inside. Not two steps in, people began rushing into the hallway – the servants from the kitchen, the tiny form of Joane, and her entourage from the parlour.

'It's you,' said his stepmother, her disappointment palpable.

'Where's Father?' asked Halley.

Joane frowned, then turned her back and walked into the kitchen.

'Where is my father?' Halley demanded.

A bumptious-looking man with greying temples squared his stance as if to prevent Halley's encroachment. It was Robert Cleeter. Halley recognised his pompous behaviour at once.

'Your father didn't come home last night,' said Cleeter, crossing his arms.

Behind him, wearing his usual apologetic expression, was young Joseph Chomat, Joane's nephew.

'Have you both moved in? I always seem to find you here nowadays.' Halley sidestepped Cleeter, taking satisfaction in the bump of their shoulders, and followed Joane into the kitchen. She was pacing around the table, biting her nails.

'Did Father go to Kent as he planned?'

'How do you know about that?'

'I'm his son.'

They stared at each other suspiciously as the others filed in around them.

'Yes, he went to Kent. Why would he not come home? We must send people to find him. Where would he go? Let's offer a reward.' Joane's voice became shrill.

Everybody started talking at once.

'Joane!' Halley's sharp cry silenced the room and turned their heads in his direction.

'He is not yet a day late. He may be trapped in the weather.'

'Look around you,' said Cleeter. 'The thaw has started.'

'Then he may have decided to stay in Kent to see another property.'

'When did your father ever do anything on a whim?' said Joane. 'Everything has to be planned out in advance and stuck to. Now, in the space of a week, he's selling the business, moving to Kent, not coming home. Nothing discussed, nothing planned . . .'

Halley knew she was right. His insides were churning. 'Let me make some discreet inquiries.'

'Discreet inquiries?' she mocked. 'I want everyone to know my husband is missing.'

'Respect your stepmother's wishes,' said Cleeter.

Halley locked his gaze on Cleeter. 'This is a family matter.'

Joane studied him with narrow eyes. 'You know more than you're saying, don't you? You were in on this whole thing, weren't you?'

Not even her extravagant silk dress could bolster her tiny frame. She was almost childlike, and not that much older than he was.

'I just want to find my father.'

'He's in trouble, isn't he?'

'You must stay calm. I will make inquiries,' he said as evenly as he could.

'Look . . . Edmond' – how awkward his name sounded on her lips – 'I know we've had our differences, but I'm asking you, pleading with you, share what you know. Let us find him. We both love him.'

Halley looked into her eyes. Again the explanation sprang to his lips but he dared not voice the words. He could not confide in a woman he loathed when he had not yet told his own wife.

He looked away. 'I know nothing.'

'You coward!' She squeezed her eyes shut, but not before a tear escaped.

'Leave well alone, Joane.' He made for the door. 'I will handle this.'

Her voice halted him. 'I won't forget this, you know.'

Halley slammed the door, more angry with himself than with Joane.

16

Hooke cast his eyes over the half-finished church. It was tall, squeezed between two dwellings with narrow walkways on either side. As usual a number of passers-by had stopped to watch the work, commenting variously about how quickly or slowly they judged the construction to be proceeding. Today the workers were slotting in the window frames so that the stonemasons could continue building around them tomorrow.

Hooke allowed himself a small glow of pride. It was going to look magnificent. It was a church such as this that, as a boy nurturing an ambition to become a vicar, he had imagined would be his. Cromwell's years of Puritanism had put paid to that plan, yet Hooke still wondered what his life would have been like had he been ordained.

The noise of a carriage coming to rest drew his attention. The lofty figure of Wren, all cuffs and buttons, uncurled from the door.

'I hadn't expected to see you today, Kit.'

'Hungry?'

The emetic Hooke had taken that morning had prepared him for work but left him hollow. Once at the trestle tables of the dining-house, noisy with the chatter of other patrons, he forced himself to eat sparingly. He was well aware of the consequences of overloading his stomach, particularly of late since the strange episode outside the theatre, but the smell of the roasting pig over the fire did not make his restraint easy.

'Are you aware of the date?' There was an apprehensive tone in Wren's question.

'Of course,' said Hooke.

'So you remember my offer of two months ago?'

'You're talking about gravity.' Hooke dropped a greasy morsel on to his plate and wiped his fingers down his jacket.

'If you have the proof, show us all,' said Wren. 'Let us enter it into the register at the next meeting and the discovery will be yours. All too often in the past you have taken steps along a path but failed to complete the journey. Then, when you're beaten to the conclusion, you have called for attention.'

'I don't recognise myself in that description.'

'What about when that Dutch chap – oh, what was his name? Christiaan . . . Huygens, yes Huygens – when he visited and brought the watch he had invented? You claimed to have already built one, but were unable to produce the working model to show us. Then you claimed to have invented the reflecting telescope when Newton presented his. Don't make the same mistake with gravity. I hear things, Robert. Some of the Fellows consider you too full of bluster.'

'They treat me as a common workman, dropping their requests for experiments on me at the last moment and complaining if I don't have the perfect apparatus at the very next meeting. Sometimes they only give me twenty-four hours to prepare. Yet I'm never congratulated on my successes, only chastised for my failures. I should be recognised for my contributions.' Hooke's stomach began to churn. He looked around the room at the other diners, resenting their easy conversation and laughter.

'Robert, you're a master of experimentation and no one doubts that, but you're not a mathematician. Neither am I. Mr Halley is, and Mr Newton even more so. We should be content with our stations.'

Hooke looked into Wren's wide-set eyes. 'Why do we place such disproportionate value on mathematical descriptions?

The true insight comes at the beginning of the investigation, in the observation and the reasoning. The calculation is but the finish of the task. My church stands outside, yet I have not dirtied my hands once in its construction. Mathematicians are the workmen of philosophers, but we insist on treating them as architects.'

Wren shook his head. 'The new mathematics is the way of proving things beyond doubt. Philosophy and mathematics are drawing closer, so close that I think we may soon see a time when they will become indistinguishable.'

'Then there is no room for me in such a future.' Hooke shoved his half-finished meal away.

'Look around you, Robert; London is your testimony. You have been involved in all aspects of the rebuilding. You don't need anything else.'

'You will be remembered for London, not me. You designed most of the buildings; I merely surveyed the land they stand on. Once more, I am cast in the shadows.'

Wren sighed. 'Oh Robert, you must cheer up, or I fear this self-pity will destroy you.'

17

Islington Village

Mary's soft touch on Halley's shoulder broke his reverie. The folds of her dress rustled against his chair and he reached up to take her hand.

'What are you working on?' she asked.

'Oh, just pushing equations around a page, getting nowhere. I remember Father would look at my equations and say, "They might as well be the hieroglyphs of Egypt." I tried to tell him: each letter represents a measurable quantity. Say *p* was the time the planet takes to orbit the Sun and *a* was a measure of the planet's distance from the Sun. But he wouldn't listen. He thought that to seek out the rules of the cosmos was threatening.

'I asked him once if he'd ever wondered why the planets are arranged as they are. You know what he said? "They are as God intended them to be."' Halley imitated his father's voice, elevating the pomposity. '"Don't look at me like that, young man. Mankind has believed in Jesus Christ for more than a millennium and a half. What makes you think you know any better?"'

Mary bent to kiss Halley. 'Well, it's a good question.'

'You know, Joane's got half of London talking about him with her stupid reward – one hundred pounds. She has no shame. We shouldn't be making a spectacle of his disappearance.'

There was a sharp knock on the front door.

'Are we expecting anyone?'

Mary shook her head.

When they reached the landing, the sound of conversation was drifting upwards. One voice was Mrs Fletcher's, their housekeeper; the other was a man's, young and apologetic. *Chomat!* Halley ran down the stairs, Mary a few paces behind him.

Joseph Chomat was waiting, face downcast, hat in hand. He looked up. 'Edmond. . .'

'You've found him?'

Joane's nephew looked despairing. 'I have bad news.'

Mary grasped Halley's hand.

'Where is he?' Halley said.

'Kent. I'm sorry – his body was washed up on the mudflats near Rochester.'

'Drowned?' Halley looked around as if lost. 'Where are my manners? Do please come through; let us offer you some refreshment.'

They seated themselves in the front room, where the dappled spring light was falling through the bay window. 'Who found him?' said Halley.

'A local boy.'

'How is Joane?'

Chomat shook his head. 'She's refusing to pay the reward, claiming that it was intended only if your father were found alive. The local authorities have backed the boy and it looks as if the matter will go to court.'

'The problem is of her own making,' Halley mumbled.

'Edmond, there's something else I have to tell you.'

'The coroner has declared that your father was murdered.'

'I must see him,' said Halley.

'The coroner?'

'My father.'

The mortuary was a long, cold tunnel with a row of skylights.

I've seen a dead body before, Halley told himself as he walked down the final few steps. His mother, his brother and his sister had all been laid out before burial. A sailor had succumbed to fever on the journey home from Saint Helena. The body had looked peaceful despite the agony of the man's final days.

A man was sharpening a long knife on the side of a stone buttress. He set it down, wiped his hands on his leather apron and spoke in a slow voice. 'Edmond Halley? They told me you were coming. Follow me.'

Each side of the tunnel was lined with wooden benches. Here and there, seemingly in random order, was the outline of a figure covered in a sheet. Some were big, some small; one was clearly an infant. Halley glanced back to remind himself of how far he was from the exit.

'We brought him in night before last,' the man was saying. 'We only transport them at night, when it's cooler.' He stooped to pick up a bucket. 'You might need this.'

Halley dumbly accepted, already feeling queasy.

They approached a bench where a large body was visible under a sheet. The mortician peeled back the covering to reveal the face and chest. Halley gagged.

'He's been in the water a long time,' said the mortician.

Halley could recognise nothing of his father in that grey mass of decay and disfigurement. Broken bone and teeth, withered muscle and flesh. One eye was gone; whether gouged before death or lost to the fishes, Halley could not tell. There were dark markings across the ruined flesh of the upper torso.

'Are those bruises?'

'He met a violent death.'

Halley nodded. 'Do you have his clothing? His possessions?'

'This is how they found him.'

'Naked?'

'Not quite.' The man replaced the sheet over the face, then lifted it from the other end, revealing the feet.

Halley choked back tears. His father's feet were still stuck in those infernal leather shoes.

Joane was dressed in black, which made her seem even smaller than usual. She was sitting beside the maw of the empty fire in the gloomy back room, staring at the wainscoting. Cleeter towered behind her chair, dressed in a sombre grey waistcoat and jacket.

Halley fidgeted. Their attire made him feel inadequate. He was wearing a simple black armband, but his jacket was fastened by a dozen gold buttons that seemed inappropriately cheerful.

Five minutes had passed since he had been shown in, and she had yet to look up.

'I am sorry, Joane,' Halley said. 'I know how you must feel. I assume that he is to be buried in the family grave.' He stopped himself from saying, 'with Mother'.

'I have made the arrangements with the Vicar of Barking,' she murmured.

Halley was thinking that his best course of action would be to leave, but then she looked up at him.

'I knew about the Earl of Essex. Your father was terrified. Did you really think he wouldn't talk to me – his wife – about that?'

But that had been our secret, Halley thought; he and his father had agreed to carry the burden alone.

'I gave you the chance to confide,' said Joane.

'If you knew the truth, you should have known to keep quiet. It may not be over.'

She tutted. 'Of course it's over, Edmond. Haven't you heard? Braddon and Speke have been arrested, tried and found guilty of spreading derogatory rumours. They're in prison. And now your father, their principal witness, is dead. I would say that closes the matter, wouldn't you?'

'Then we have no more to say,' he said stiffly. 'I will be in touch about taking over my father's estate.'

'It's not as simple as that,' she said. 'You'll know soon enough, so I might as well tell you now. Your father died intestate. I'm suing you for control of his properties and wealth.'

'My father's properties are mine. I'm his son.'

Cleeter stood gawping behind Joane like some ridiculous butler.

She stood up. 'Since when have you shown any interest in your father's business? Always too busy careering around trying to impress the greybeards at that Royal Society of yours. You forfeited your rights to your father's estate a long time ago with your indifference. I'm the one who's helped him these past ten years.'

'You will get this house and your share of his personal wealth. With the sale of the soapworks that will be enough.'

'There's nothing left from the soapworks. Your father had just mortgaged the place to buy new vats. He sold everything at exactly the wrong time. You'll get something of his wealth and properties – I'm not as callous as you think – but I'm not handing everything to you.'

Halley turned to go.

'That's right,' she said in a brittle voice, 'do what you always do. Run away. Pretend it's not happening. How will you ever amount to anything? Your father mollycoddled you when he should have made you work. Well, Edmond, it's time you grew up.'

That night, cradling a goblet of wine, Halley watched the Earth's celestial neighbour rise. He should have been out there, under the Moon, making his observations, but Joane's voice rattled in his head.

Time to grow up.

He tried to dismiss her words, but they had lodged inside him. He threw the rest of the wine down his throat and turned away from the night sky.

18

Halley spurred his horse faster, relishing the breeze on his face and the flash of the hedgerows as he thundered along the Great North Road. The drumming of the hooves on the dried earth became one with the pounding of his heart as he galloped past the lumbering coaches sharing the road.

Now and again a cluster of whitewashed cottages sped past; a few villagers looked up to see who was in such a tearing hurry. Perhaps they expected to see the red livery of a King's messenger. Instead, they saw Halley's wind-tousled hair and billowing shirt. He wanted to laugh out loud, and it struck him that this was how he used to feel all the time. It had been four months since his father's funeral. The world had continued through its orbit, and the air was again filled with birdsong and warmth.

He was riding so hard that he missed the fork to Sawtry, where the reluctant rector and the long-overdue rent were waiting. Drawing the stallion to a halt, he circled back to the split in the road. Sure enough, he had taken the route where someone had used a poker to burn the word 'Cambridge' into an arrow of wood.

On the grassy verge startled geese flapped furiously to escape the horse and rider. One by one they rose clumsily into the air, becoming more graceful as they gained height. The sight reminded Halley of Hooke, who had once said that the grace of a high-flying bird must surely be attributable to the decrease of gravity with altitude.

Hooke. The deadline for his proof had long since passed.

Halley looked thoughtfully at the sign to Cambridge. The horse was growing impatient beneath him.

'Hah!' he shouted, slapping the reins.

The Rector of Sawtry could wait.

He galloped off towards Cambridge and Newton.

Summer air clogged the Cambridge streets. Halley found lodgings at the first half-decent inn he came across, then set off into the maze of narrow walkways overhung by rickety timber buildings. Conscious of the stares drawn by his tailored jacket, he increased his pace.

Through the slivers between the buildings Halley at last spied the four towers of Trinity's gatehouse. He emerged from the alleys and crossed the open concourse.

Someone jumped from the shadows of the college garden's wall. A malnourished lad dressed in sackcloth, with a shrivelled right arm, confronted him and nodded a greeting.

'You startled me,' said Halley, making to step around him.

'Who you here to see?'

'Mr Newton. Not that it's any of your business.'

'Aye, he's in. First room on the right, when you get through to the quad.' There was something in the way he emphasised the word 'quad' that indicated he had no idea what a quad was. Probably just something he had overheard. Yet there was sincerity in the boy's eyes, and a painful neglect, evident in his scuffed and dirty knees, that tugged at Halley. He reached into a pouch and tossed the boy a coin.

'Go through the small one. They only open the big one for the Master,' called the boy, holding his prize.

'Thank you,' said Halley as he pushed open the gate and stepped into the coolness of the stone entrance.

A steward looked him up and down and asked him to state his business.

'Edmond Halley, astronomer and Fellow of the Royal Society, to see Mr Newton.'

'Wait here. But don't get your hopes up. Mr Newton doesn't have many visitors.' With a quizzical look, the steward left.

Dons in their black gowns came and went. When the steward returned, he seemed perplexed. 'He says to go round, to the first –'

'Door on the right? Thank you, I've been here before.'

Halley stepped from the shadow of the gateway into the full sunlight of the path that ran around the quadrangle. The door to Newton's room was closed. Halley knocked and waited. The door opened a crack and a single bloodshot eye squinted as him. It was not Newton's.

'Yes?' its owner inquired, as if Halley's arrival was entirely unexpected.

'Is Mr Newton in residence? I'm Edmond Halley.'

The door opened enough for the man to bend himself into a bow. He spoke in a hoarse whisper. 'I'm Humphrey, sir, Mr Newton's assistant. He bids you come in and wait.'

Halley stepped in from the brightness, straight into a heavy swathe of dark fabric draped across the doorway.

'Wait! Light will ruin the experiment,' explained the man, closing the door behind Halley with a click of the metal latch. 'There, you can go through now.'

Halley parted the curtain and stepped into a room where illumination was a precious commodity. Apart from the dust motes dancing in the occasional golden needle of light from the window shutters, everything was silhouettes and shadows. As his eyes adjusted to the gloom, he could discern shelves laden with books, a desk, a couple of chairs arranged haphazardly, and in the far corner a pair of none-too-comfortable-looking beds. Piles of papers and books covered the floor.

Newton was at work in a separate space, visible through a hole in the wall. His lanky frame was crouched over a small tin furnace that, despite its size, had raised the temperature of the room to sweltering point. A rubicund glow cast him in a demonic hue. He did not look around but continued to stir a potion in a crucible. Metallic odours stung Halley's eyes and made it difficult for him to breathe.

Humphrey headed into the wooden outbuilding. He gently placed his hands on Newton's shoulders and lowered himself to match Newton's pose, sliding his hand along Newton's outstretched arm to take over the stirring. It was a practised move that ensured the liquid never broke its motion.

Newton rose, pushed up his baggy shirtsleeves and ducked through into the main room. 'Mr Halley,' he said, dabbing at his sweaty brow, 'what brings you to Cambridge?'

'Diversion from my lamentable daily routine of landlord.'

'Of course. My condolences on your loss and my sympathy for the predicament you now find yourself in. Having to share an estate cannot be an easy thing.' Newton lowered himself into a chair and indicated the other for Halley.

'The news of my settlement has reached Cambridge?'

'I have my eyes and ears in the capital,' said Newton, as if trading a confidence.

Newton had aged considerably, thought Halley; his hair was now entirely grey and sat in wiry bushes about his head. A ragged fringe had been chopped into the front. But his eyes were the same, wide open, with a gaze that was difficult to hold for long. Perhaps it was Newton's superior decade in age, or the lengthy nose he tended to look down, that made Halley uncomfortable.

'You have spent too much time with the quicksilver,' he joked nervously.

Newton raised his chin.

'I mean the colour of your hair.'

'Am I totally silver now?' Newton craned his neck, trying to see his own hair.

Halley nodded slowly before Newton fixed him with an unblinking stare. 'Tell me, what does London think of my pursuits?'

'What do your London eyes and ears tell you?'

'They never tell me about myself, only others.'

'Very well, I will not lie to you.' Halley steeled himself. 'Alchemy's considered to be a waste of your mathematical talents.'

'If you listen to others, most of my career has been a waste of time. Once it was the mathematics that was the waste, then optics, now the chemical arts. Yet it was in that furnace that I made the alloy that I polished into the telescope's mirror.'

Halley glanced over at Newton's assistant, who was continuing to stir as if in a trance. 'What are you making?'

'I cannot tell you. Unlike the Royal Society, I don't believe that all knowledge needs to be widely known. The followers of Pythagoras were a secret cult. Only after their work was plundered did the rest of the world learn of its existence. You of all people should appreciate the wisdom of keeping some beliefs private.'

Was he referring to religion now, or that stupid business with Mrs Hevelius? 'Another snippet from your London eyes and ears?'

'There will be tough times ahead for atheists,' said Newton matter-of-factly.

Halley inclined his head as if accepting advice, then said, 'Mr Newton, in London the problem of celestial motion vexes us all. We would like to know what shape an orbit would be for a planet held in place by a force that drops with the inverse square of distance.'

'An ellipse,' said Newton without hesitation.

Astonishment blossomed in Halley. 'How, sir, do you know that so certainly?'

'I have calculated it.'

'Calculated?'

'Yes, I calculated it some years ago. I have the paper here.' Newton rose from his chair and began to rummage through the bookshelves.

Halley could not bear to stay in his seat, but sensed that any attempt to help would not be welcome. His heart quickened when Newton pulled a sheet of parchment from a book and examined it closely. The don then slid the page back into the book and replaced it on the shelf.

'I cannot find it,' he said, without looking round. 'I will have to perform the calculation again and forward it to you.'

'Can you remember the steps you took in the solution?'

'Of course.'

Halley waited, expecting Newton to at least outline the stages, but Newton returned to his seat. 'Who works on this problem in London?'

'Sir Christopher, Mr Hooke and myself have all given it thought.'

'Mr Hooke? How far has he got with the problem?'

'He has ideas, as we all do, but he is no nearer a solution than the rest of us. I fear his algebra will let him down. Unlike yours.'

'Of course, some say that algebra is the language of God.'

'Then you must understand Him more than any man alive.'

'I was born on Christmas Day. That must count for something.'

Halley laughed, but a sharp look from Newton silenced him.

'Mr Halley, I will send you the calculation on one condition: that you keep it to yourself. You must not show it to anyone, especially not to Mr Hooke. I have found in the past that philosophy is a most litigious lady, with Mr Hooke her most bombastic admirer.'

'I assure you that a proof of the elliptical orbits of the planets – if you did decide to make it public – would receive nothing but open-hearted praise. For my part, I'm content to receive such knowledge under any conditions that you see fit to impose. You can rest assured that I will follow them to the letter.'

Newton pursed his lips. 'Then I will do this on account of our acquaintance. It will be my final gift to philosophy. But, I emphasise, it must remain strictly between the two of us.'

'Granted.'

'I am adamant.'

Halley met Newton's cold gaze. 'So am I.'

Newton's fingers were still smarting from Halley's parting handshake some minutes after the young man had bounded from the room. He was thankful that Halley had known better than to press him for a timetable. Even so, just agreeing to help rankled. He did not have the time.

Massaging his hand, he returned to the bookcase and drew out the sheet of paper again. Surrounded by his scribbled mathematics was his diagram of an ellipse, with intersecting lines cutting it into quarters. He had also drawn a couple of diagonal lines across the ellipse.

Unlike the first time he had contemplated the problem – sitting in the quiet of Woolsthorpe, having fled Cambridge during the plague summer – his breakthrough had come just a few years ago when he turned the problem around. Instead of asking what shape of orbit an inverse square force law created, he investigated what force law would create an ellipse.

Working the problem backwards like that had driven him further through the mathematics than before, as if he had just found a path through an overgrown coppice. But the path had been suddenly blocked: the equations had failed to simplify even though every sense within him told him they should work out. He had nearly confessed as much to Halley, but had stopped himself when he realised that to do so would place him on the same level as the London philosophers – the same level as Hooke.

He stared at the page; there must be an error on it somewhere. Perhaps he had confused some of the lines on the sketch. Loath as he was to admit it, drawing was not his strong point. He would check it all again, but not now. He turned the paper face-down and headed for the furnace.

This particular experiment had begun twelve days ago, when he and Humphrey had boarded up the windows and begun the ritual grinding of the limestone. They had lit the furnace and mixed the lime with sulphur and mercury, all in precise proportion. Then they began the continuous stirring that joined the alchemist to the alchemy itself.

Newton slid his hand along Humphrey's lean forearm, feeling the life within him and gaining the precise motion of the stirring from the flexing of the young man's muscles. His fingers found the wooden spatula and Humphrey slipped away, back into the main apartment.

The steely liquid should have been engrossing, urging Newton's mind back to its former neutral state, but in that circulating fluid he saw nothing but orbits. Figures from the page on the bookcase crowded into his head, and symbols too. He began to imagine how the bulk of a planet might affect its motion. What about its distance from another object? And the size of that other object? All of these quantities must define how one body pulls on another. But how precisely did that happen?

Time and again Newton chased away the dancing symbols, only for them to flash back into his mind. He shook his head, then pinched the bridge of his nose so tightly that pain lanced across his cheeks. When he returned his attention to the furnace, he saw that the uniform circulation had broken into eddies.

The furnace transferred its fire into Newton. He stood up, raised his booted foot, and sent the ruined concoction cascading across the floor with a heavy clatter.

Humphrey rushed back into the shed, stammering, 'I'll get a bucket and mop.'

'Leave it,' growled Newton. 'Fetch me quills and ink. I must calculate.'

19
London

Hooke fussed about, arranging chairs, yanking them into place and raising tiny puffs of dust from the faded rugs. He opened the folded screens that masked the messiest areas of his apartment and placed the Society's mace on the table.

The Fellows began to arrive, including, Hooke was relieved to see, the Society's new president, Samuel Pepys. Newly returned both to England and the heart of the monarchy, Pepys was as rotund as Halley was lean. Both men were chatting earnestly with Wren.

Curiosity burning, Hooke busied himself repositioning a few chairs, edging himself closer to his visitors. As he did so, Halley excused himself and disappeared deeper into the room. Hooke used this as cover and slid closer still.

'I admire your fortitude,' Wren was saying. 'I've been hearing about the gale. You're the talk of the coffee-shops.'

'We just couldn't get out of it. Went on for days. Howling, pitching. We took down all the sails and just let the sea have us. At one point I thought we were going to have to lash ourselves down. I still sometimes see the waves in my dreams, tall as mountains . . . Still, better those memories than ones of imprisonment.'

Hooke had heard the stories too. By all accounts, Pepys was something of a hero. He had evacuated the colony at Tangiers and supervised the blasting of the British fortifications into dust so that no enemy could occupy them. The service had finally erased the smears of Catholicism and accusations of leaking naval secrets to the French that had seen him briefly in the Tower.

'Strange to think that destroying an English strategic asset could lead you back into favour,' Wren chortled, but Pepys looked at him in all seriousness.

'When royal whim is involved, one needs to do whatever it takes. Besides, the King couldn't afford to defend it.'

'You're absolutely right. If someone doesn't stop the counterfeiters, not only will the King go bankrupt, so will the country. But that's a problem for the future, eh? Right now, congratulations on your new appointments, both here and at the Admiralty.' Wren bowed his head.

Hooke glanced at the clock, wondering whether it was time to call the meeting to order. An icicle pierced his stomach: next to the tall case with its weights and pendulum, Grace was talking to Halley. She was punctuating her words with smiles and glances through her eyelashes while he leaned against the wall, drinking in her flirtatious behaviour.

Hooke made for them at once.

'Must all the Fellows be so old?' Halley was asking her in conspiratorial tones.

Hooke cleared his throat.

'Uncle,' – Grace reached past Halley to where her latest creation was wrapped around a mannequin – 'I was just collecting your coat to finish the alterations.'

She carried the woollen garment out of the room but took a final peep at Halley from the doorway. The gesture was not lost on Hooke, who glared up at the young astronomer.

'Why do I sense a rebuke is about to be delivered?' said his victim.

Hooke wagged a crooked finger. 'I have warned you before.'

'We were only talking.'

'Do not bother Grace again. She's out of your reach.'

'Rest assured, I am reaching for nothing. Not that it is any of your business, but I'm to be a father. At least I hope so. Mary and I have decided that it's time to start a family.'

Hooke looked away. 'It would be best if you just sat down,' said Halley.

'Ah, there you are, Mr Halley!' The new voice was not familiar to Hooke. A small man in a ridiculously large hat was coming towards them.

'Mr Paget,' Halley greeted him, 'what brings you down from Cambridge?'

'This meeting, of course.' Paget turned his back on Hooke, who loitered. 'I'm so relieved to find you, Mr Halley. I'm under instructions to give this to you, and to you alone. It's from Mr Newton.'

Hooke peered round Paget's shoulder, catching Halley's sideways glance full-on. The astronomer took the proffered envelope, held it as if not quite sure what to do with it, then broke the wax seal.

Hooke strained to see inside, bumping into Paget in his excitement.

Halley abruptly stuffed the envelope into his jacket pocket. 'Gentlemen, I regret that I must bid you farewell.'

Before Hooke could think of anything to say, Halley had left the room.

Hooke fixed Paget with a look. 'What was in that envelope?'

'I cannot tell you, because I don't know.'

Deep down, however, Hooke already knew.

Gravity.

Once back in his Islington study, Halley excitedly scanned the manuscript. It had been months since his visit to Cambridge. He had almost given up hope.

De motu corporum in gyrum

On the motion of bodies in orbit. The title was in Newton's tight handwriting. Just nine pages in all, but each one was a dense presentation of words, mathematics and diagrams. When Halley failed to find the simple answer as to why planets follow elliptical orbits, he stifled the twinge of impatience and returned to the first page. Reading from the beginning, he still found it difficult to follow the argument. He paused for breath and began again. This time he got a little further before losing the thread and having to retrace his steps.

He fumbled in his desk drawers for paper and copied out the mathematics and the diagrams, emulating the process Newton himself had gone through. Then in a flash he realised why the text was so impenetrable.

It was because he and all the others had been hopelessly naive. Not only had they lacked the skill to answer the question, they had not even known how to frame the question correctly. Newton's paper gave no simple answer; it was an explanation of how to make the problem soluble.

This was no rough verbiage. Newton clearly stated principles and then tested the assumptions that could be built upon them, using mathematics and graphs, curves and caveats. It was intoxicating. Halley's brain pulsed with energy and surrendered to the text, allowing his mind to be manipulated into thinking as Newton had done when writing the document.

As the planet drew away from the Sun, so the force dropped and the planet slowed down. Eventually, like a ball thrown into the air, the planet fell back towards the Sun, picking up speed as it went and whipping round the Sun to fly back out into space and start the orbit off again. What everyone had suspected, so Newton had proved. Double the distance, quarter the strength: an inverse square force of gravity.

How could he possibly keep this to himself?

20
Cambridge

Newton woke from a brief night's sleep and was instantly bombarded with ideas for calculations and experiments. It was as if his brain had worked on through the darkness, waiting for him to revive so that it could force him into action.

Every morning since Halley's visit had been the same. He hardly knew where to start. All his life he had relished questions and the way his mind drove him to investigate; now he was its slave.

The moving world had become nothing but curves. From the tumble of an orange in the market, to the oscillation of a cart as it rocked out of town. Every movement that Newton saw demanded his attention.

The sky was peachy with the dawn as he strode across the quad to rap on Humphrey's door. The man appeared bleary-eyed, still in his night chemise.

'Fetch the pendulum,' said Newton.

'But it's freezing.'

'Precisely! We will be the only ones about.'

Later that morning, Halley found Newton and Humphrey bundled up and hunched over the pendulum in the college cloisters, just as the urchin had described.

The frost was thick on the grass, and the flagstones were icy in places. Halley's nose stung with the cold, and his fingers were numb despite being encased in leather gloves. He rounded the corner just as Newton began to stamp his foot. The sharp reports blended with Halley's own footfalls.

Newton shot a look in his direction. 'You've ruined the experiment.'

Halley came to a standstill. 'My apologies, I had no idea.'

Newton peered. 'Mr Halley?'

'Yes, sir.'

'Well, well. I fancied that I would see you again after sending my little paper.'

'Nine pages of brilliant mathematical insight, if I may flatter you so.' Halley reached into the satchel slung across his shoulder and waved the document. 'I lost a night's sleep reading and re-reading it.'

'Only one?' said Newton, bending back down to the apparatus.

De motu in hand, Halley addressed Newton's back. 'You have derived the planetary motions from first principles. It's the perfection of astronomy, the–'

'Not yet, it isn't.'

'You tease me, Mr Newton,' said Halley.

'Not at all.' Newton shortened the length of the pendulum string.

'I know that you wanted your ideas to remain secret, but I should like you to reconsider. They are too valuable to remain unseen.'

'No!' Newton whirled to face Halley. 'Would you say that you have described the bird if all you had was a feather?'

Halley shook his head, discomfited by the intensity of the professor's gaze.

'Well, you hold before you the feather, not the bird.'

'But you have derived Kepler's three laws from an inverse square force. What else is there to do?'

'What else indeed?' said Newton, striding off.

Humphrey spoke quietly while pretending to be busy with the pendulum. 'Mr Newton is upon this problem with every ounce of his being. I've found his meals untouched, and when I remind him he takes only a mouthful or two. If I succeed in urging him to the dining-hall, he is as likely to turn left instead of right and end up in the street. I lose him into the city, only for him to return later and continue working. He hardly sleeps. I have seen him jump from the gardens and run to his room to scribble frantically, breaking one pen after another in his haste to set down his thoughts. He's lost all his students – he talks to the bare walls about his ideas – and he has me copy his manuscripts, but I can barely make any sense of them. I fear he is in the grip of some mania and is producing nothing but gibberish.'

'Fifty-seven paces,' Newton called from the end of the cloisters, start-ling both men.

Humphrey scribbled the figure down.

Newton walked towards them, talking. 'So far, all I have done is show how the Sun pulls on the planets. But how does Jupiter pull on the Sun, or on Saturn? How does the Earth pull on the Moon, and does the Moon pull back? I have perfected nothing because I have understood nothing – I've barely glimpsed it.'

'But it is a mighty first step, can we at least agree on that?' ventured Halley. 'There are many who–'

'*De motu* is grossly incomplete – that is the end of it. You see this work as finished, but I know it is just the beginning.'

'As you wish, but let me at least describe it at the next meeting of the Society. That way the discovery can be secured in your name.'

'What about Hooke?'

'What about him? He has had his opportunity. When I look at the mathematics required in your work, it is beyond Mr Hooke – beyond all of us.'

Newton's lips narrowed in what might have been a smile. 'Very well, you may describe our meeting, but not show the document. There are certain clarifications I wish to make, certain aspects of the present paper that could lead to discussion. And I am unwilling to enter into another challenge. I will fix those problems, and then you may publish it. And I will need the comet observations you made.'

'You think you can do the comet's paths as well? I will copy them out and send them as soon as I get home. When do you think the new *De motu* will be ready?'

Newton tutted. 'Mr Halley, I must examine every aspect of what I propose. Hence today's experiment to measure the speed of sound in air, so that I may understand how disturbances move through fluids. I tell you this because you are a friend. Now, quiet, please. I'm ready.'

Humphrey set the short pendulum swinging to and fro. Newton stamped his foot as the bob reversed direction. An instant later, the echo from the other end of the cloister returned, just as the pendulum reached the other side of its swing. Newton waited and stamped again. The echo fell in perfect time with the pendulum.

'Measure the string. That's the length we need for the calculation.' Newton looked surprised to see Halley still standing there. 'Return to London, Mr Halley. Wait for me to contact you. Now that I am upon this subject, I will know the bottom of it before I publish a single word.'

21
Greenwich

Halley had always thought of the Royal Observatory as a gaudy fortress. Perched on the hill overlooking the growing naval yards of Greenwich, the building hid its eccentricity behind an unassuming red-brick facade with whitewashed turrets and curlicues.

He found the best approach was to take the steep hill at a brisk pace. As he did so on this occasion, he passed a well-dressed walking party, their dogs bounding in circles and their ladies treading carefully across the uneven path. The menfolk were pointing out the observatory, but Halley did not stop to listen to their comments. In his experience most recognised the place as where the King's astronomical work was undertaken, but few realised the magnitude of the task: the thousands of stars to be pinpointed with the modern telescopes inside, and the pages of calculations to be converted into navigational star charts.

As he entered the courtyard the true nature of the building became apparent. The observatory's main bulk was a double-height octagonal room of giant windows and roof-level balustrades. He was not the first to arrive. Two men, red-faced and sweating despite the bitter day, were carrying a pair of mattresses up the observatory's steps. Nearby two boys in blue school coats shivered, waiting their turn. One had a mop of red curls, the other lank curtains that had been chopped at shoulder length. Each carried a valise.

'Come on, lads, let's get you inside,' he said. 'You must be the new students.'

The redhead scampered up the steps.

'Assistants,' said the other pointedly.

'You'll get on very well with Mr Flamsteed,' said Halley as the boy passed him.

Halley followed, squeezing past the two workmen on their way out. Behind them was Flamsteed, dressed in a full black cassock. There was a pallor to the astronomer's face that Halley did not remember seeing before.

'Lads, allow me to introduce you to the Reverend John Flamsteed, newly ordained Rector of Burstow and the King's Astronomer,' said Halley with a flourish.

'No need for theatrics,' said Flamsteed over the students' muted greetings. 'This way, boys, we have much to prepare for tonight's eclipse.'

The lower storey of the observatory was a basic affair of dark wood and small windows. It consisted of just four rooms and a short connecting corridor. The new mattresses had been laid on the floor next to Flamsteed's bed, presumably so that he could read the Bible to the boys before they went to sleep. The bluecoats deposited their belongings, and Flamsteed ushered them into the front room where the fire grate was forbiddingly cold and dark. The redhead looked at it with a mournful expression.

'Better you get accustomed to the cold before we start the night's observing; there'll be nothing to keep us warm then.' There was a rattle in Flamsteed's lungs as he spoke.

He set them to work with a lunar map, explaining that the Earth's shadow would sweep across the Moon from the south-east, and that they were to number the major craters in the order that Earth's shadow would cross them.

'By measuring the time the shadow crosses each crater we will be able to calculate the size of the Earth and the distance to the Moon,' proffered the serious one.

'Correct, young sir.' Flamsteed smiled. 'Now, get to work.' He turned to Halley, who was waiting at the door.

'Tell me, John, did I seem that young when I first came to study with you?' asked Halley.

Flamsteed pulled a face as he pushed past. 'No.' He pointed at a mattress under the kitchen table. 'That's yours.'

Halley followed him, crouched and tested the softness, rustling the straw filling. 'Thank you, the wherryman is returning for me at dawn . . . Tell me, John, are you well?'

'A chill caught observing, that's all. Don't fuss; it will pass. There's something I want to ask you. What's happening at Cambridge? Mr Newton is up to something, and you're his confidant, I understand.'

'I have seen a confidential draft of a work.'

'And?'

Halley wrestled with breaking his oath. 'I cannot divulge anything until the next Society meeting.'

'You choose him over me, after all the tutelage I gave you?'

Halley held his hands up. 'No, John, no. Please, if I tell you, you must not spread it. Mr Newton is upon the subject of gravity. He has already shown how Kepler's laws can be derived from an inverse square law, and I believe he's now searching for more consequences of gravity.'

'So that explains why he asks me whether the moons of Jupiter follow Kepler's law.'

'And do they?'

'Yes. They can be considered as a miniature system in their own right, with Jupiter replacing the Sun and the moons replacing the planets. Of the planets, only Jupiter and Saturn occasionally disobey Kepler's laws – and then only when they draw close to each other. Mr Newton requires those observations, too. I receive new demands for astronomical data every few weeks.'

'What else is he asking for?' Halley's heart pounded with excitement.

'Last week it was the precise position of two stars in Perseus.'

The comet. It had passed through Perseus.

'His last request is unfathomable,' continued Flamsteed. 'He wants information about the tides at Greenwich. His thinking is all over the place, and I cannot keep dropping everything to respond to Mr Newton. I have a star catalogue to complete.'

They were interrupted by the red-haired boy, who popped his head round the door. 'We've finished the numbering, sir.'

The Moon, full and ripe, was visible through the star room's windows, its dark markings almost obscured by the brightness of the rest of its surface.

Cold air tumbled into the room through the open panes, wrapping them in icy fingers that provoked a coughing fit in Flamsteed. The telescope they were to use that night was a spindly affair, at least a dozen feet in length but a mere four inches in diameter. Halley gripped the end of the instrument and climbed a ladder that was secured across the open window. He rested the telescope on one of the rungs so that it pointed to the Moon as Flamsteed secured the base into a wooden cradle and dipped his head to check the alignment.

The boys took up their positions near the life-sized portraits of King Charles and the Duke of York that concealed a pair of pendulum clocks. Only the pale faces of the clocks were visible, and their ticking was muffled behind the panelling.

Halley descended the ladder, winked at the bluecoats and returned to Flamsteed and the telescope. Even to the naked eye there was a black nibble at the lower side of the Moon.

'First contact of the umbra. Time?' called Flamsteed.

The redhead called out the position of the hands, and the other carefully noted them down.

Every time the sharp darkness kissed the lip of a crater, Flamsteed called out its number on the boys' chart and barked for the time.

After some fifteen minutes, Flamsteed turned away from the eyepiece. 'Small break,' he announced. 'Next crater in four minutes.' Massaging his fingers, he blew into them and returned to his vigil.

'Will you miss this place?' asked Halley.

Flamsteed did not look away from the telescope. 'What do you mean?'

'When you move to Burstow?'

'What makes you think I am moving?'

'You're the new Rector of Burstow. You'll live in your parish. How can you continue your duties as King's Astronomer?'

Flamsteed rocked backwards. 'Why, you insolent pup! Are you seeking to replace me here?'

'Of course not! Well . . . all right, yes, I confess the thought did cross my mind. I need a job. Mary has a child on the way. My stepmother has stolen half of my inheritance–'

'I have no intention of giving up this place until the King's catalogue is complete or the Good Lord calls me to Heaven. Do I make myself clear?' fumed Flamsteed.

'I apologise,' muttered Halley, cursing himself for being so ill-mannered.

Flamsteed turned back to the eyepiece. 'Time!' he barked, more harshly than before.

As the half-hearted winter dawn inched across sky, the clang of church bells drifted down the icy Thames. What started as a single toll grew in urgency until it seemed as if every bell in London was ringing.

Halley stumbled from his makeshift bed and came face to face with the King's Astronomer, mid-coughing fit.

'Something's wrong,' said Halley.

Flamsteed pulled open the shutters and they peered out.

There were ships on the bend of the river, their white sails as small as handkerchiefs.

'Please forgive me for my clumsiness yesterday, John. Heavens above, cannot you see my wanting your job as a compliment?'

Flamsteed kept his back to Halley. 'And now you compound an insult with a blasphemy. Your time at sea has done you no good, young man. You swear like a sea captain.'

Halley spotted a figure on the steep path. It was the wherryman, charging up the hill swinging his arms.

'I'll see myself out,' said Halley.

When he opened the front door the wherryman was bent forward, resting his hands on his knees. He nodded towards the river. 'It's . . .' he said between giant gulps, ' . . . it's the King.'

The urgent clanging of the bells jarred in Halley's skull as he slithered around the wharfs. As he became caught up in the crowd of people hurrying towards the churches, a familiar crooked figure drew his attention.

'Robert!' Halley made after the shuffling form, moving as fast as the frosty pavement would allow.

Hooke looked round and grasped at Halley's sleeve. 'The King needs our prayers; he's gravely ill.'

'Robert, you know I don't–'

'Even you can do this for the King.'

'It's not the King I find it difficult to believe in.'

Halley was swept through the stone entrance by the flow of people and found himself beneath the high vaults. He followed Hooke into the pews, and soon the vicar was rousing the crowd with volume, leaning over his pulpit, the whites of his eyes rolling, assuring them that the power of prayer would save Charles. The people squared their shoulders, straightened their backs and looked ready to march to war. Then they bowed their heads to pray.

Later, as the church emptied, a ripple spread through the crowd. Trying to catch the words, Halley looked at Hooke, whose face was aghast. The ripple passed again, moving from lip to lip. This time it was explicit: the King was dead.

One of the choirboys began to cry.

A gruff shout went up from somewhere in the crowd: 'Long live the King!'

And the ripple became a single thunderous wave of repetition. Then the crowd fell silent, crusaders robbed of their crusade.

The two men repaired to a tavern near the river. It was unusually busy for the time of day, and they had an odd sense of being out of place. Much of Hooke's ale ended up spilled down his front as first one patron then another knocked into him.

Halley looked down at his friend. 'Come, let us find some breathing space. There's a room at the back.'

They tunnelled through the crowd; the nearby conversations were full of poison and regicide. 'It's the only explanation for the speed of it,' said a pockmarked man, bringing his fist down on the table to emphasise his point.

'No,' protested his companion, 'it's those experiments they say he did, every day, they say, working with the quicksilver. And did you hear? He converted back to Rome just before he died but choked on the holy water. Seems the Almighty doesn't like a hypocrite either.'

Halley led Hooke to a tiny, yellow-stained room which so far had been overlooked by the other patrons. They slid on to a bench.

Hooke clanked his pewter tankard on to the table. 'You spent the night at Greenwich, you say?'

'Indeed, observing the lunar eclipse. But I have fearfully upset John.' He explained how he had inquired about his job.

'I never thought I would laugh on the day the King died,' chuckled Hooke.

'Laugh all you want, but the star charts are urgent. I do think they could be tackled with a little more haste. Robert? Robert!'

Hooke's eyes flickered and his body rolled from the seat, sending the beer splashing over the table and on to the floor. Halley leapt to his feet but he could not avoid the dousing, nor could he prevent Hooke's collapse. The Gresham Professor hit the ground with a thud. Halley rolled the inert form over and brushed the sawdust from his face. There was a livid mark already visible on Hooke's forehead. His mouth was moving, emitting small noises, and his eyes fluttered somewhere between lucidity and unconsciousness.

'Come on, Robert. Let's get you up.' Halley took the man by the armpits and heaved, but there was no strength in Hooke's legs. His head lolled about as Halley hoisted him over his shoulder and manoeuvred through the inn.

'And him a gentleman, too,' slurred the pockmarked drunkard as they passed.

'Grace?' Halley called, reaching the apartment's front door. 'Come quickly!'

She appeared from the kitchen, wiping her hands on a cloth, and drew in a sharp breath at the sight. 'Not again?'

'This has happened before?'

She nodded, looking at her uncle's face. 'At times, since he was assaulted. It's getting worse. Let's get him to his bed. He'll be fine after a sleep.'

She shut the door and led the way upstairs to a small, fusty bedroom. Halley lowered Hooke on to the bed. Grace loosened her uncle's clothing and removed his shoes, then covered him in blankets.

Downstairs, Grace offered Halley a glass of wine.

The aromatic smell teased Halley's nose. 'Cinnamon,' he said with some envy. 'Expensive stuff.'

'Uncle lets me choose.' She poured herself a drink and stepped closer to Halley. 'Your good health, sir.' He toasted her in return, guiltily enjoying her proximity and those dark, soulful eyes. He felt breathless.

'Uncle says you're to be a father.'

Halley nodded, embarrassed that she should know.

'The confinement of a wife takes its toll on a man.' She took a slow drink, keeping her eyes on his.

'I have to go.'

'Why? Uncle will be asleep for hours, nothing can rouse him after one of these bouts.'

The soft pleading in her voice paralysed Halley. It had been so long since Mary and he . . . Grace seemed so young, so vulnerable. He wanted to take her in his arms to comfort her, just to comfort her. Her lips were glistening, he noticed, and he sensed his mouth parting in response.

Grace pressed herself against him, lifting her face to his.

Halley closed his eyes as their lips melted together.

Then he broke away roughly and fled.

22
Cambridge

'I was wrong to call it an inherent force. Whatever the property is that resists change in the motion of an object, it's not a force. A force can only exist when an object is being accelerated. What are the three states of motion? Stillness, uniform velocity and acceleration. Only in the last case is there a force acting. That means that bodies at rest and those moving because they have already been pushed are equivalent; they have no force acting on them.'

Newton paced the room, raking his hands through his hair. There were no students on the benches to listen to his lecture, but he did not let it trouble him.

A letter from Halley had arrived that morning, explaining that discoveries could not be entered into the Royal Society's statute book without a paper being presented. The young man was clearly fretting about securing Newton's priority, but Newton knew that no one else was having these thoughts.

Even the word 'thoughts' was inaccurate. It implied some kind of order to the flow. Newton's head was a jumble of voices, all shouting at him at once. If he could isolate a particular voice, he could channel it to his fingers and scribble furiously to immortalise the inspiration. Afterwards, he would collapse and snatch a few hours of unquiet sleep. Yet most of the time he was paralysed by the cacophony in his head. There were too many new problems to be solved. The world was in constant motion.

He began talking again but not coherently, just making sounds that echoed the equally incoherent picture of nature trying to burst from inside him. He fell silent only when he realised that there was someone in the room, a silhouette against the wall.

He squinted. It was a tiny woman. *A woman?* How could there be a woman in the college?

He froze.

– *Mother?*

– *You always did think too much, Isaac.*

A whinnying horse arrested his attention. Outside, a small procession was heading up the street: half a dozen men in tricorn hats and red robes of varying ornamentation, some wearing golden medallions. The procession stopped. He heard the clanging of a handbell. The mayor drew himself upright in the saddle. 'Today, the City of Cambridge proclaims the Duke of York, King James II, the supreme leader of all England. We remain now and for ever his most loyal subjects. Long live the King!'

Those on the street gave a cheer of affirmation.

The floor was rippling, the walls undulating around Newton.

James. Catholic. Devil incarnate.

He stumbled towards the back of the room, seeking his mother.

The room was empty.

He flailed his arms, desperate for a handhold. There was no purchase to be had.

'Mother . . .'

He crashed to the floor, his gown ballooning and settling over him as a shroud might over a dead body.

Part II
Distance

23
Cripplegate
1686

James II wasted no time in tightening his grip on the country. He appointed Catholics to positions of power in his Privy Council and the army, and when Parliament objected to this violation of the Test Act he prorogued it, sending the gentlemen home to their constituencies to mutter and moan. He bolstered the rank and file of the army with Irish Catholics and stationed it to the west of London – a sprawling, chaotic mass of tents and campfires.

Halley saw it one day from the road. Pennants fluttered in the breeze and a carnival atmosphere induced the soldiers to drink and cavort with the nearby caravans of whores and tinkers, all too eager to take the men's money.

The sight kindled memories of his father's stories of the civil war: the lootings, the rapes, the burnings. A standing army was an expensive luxury for a monarch. Only if it were well fed, well paid and well serviced was the rest of society safe.

Halley had made the mistake of telling Mary what he had seen, and every night after that, when she arose to feed their newborn, she imagined the soldiers creeping into the silent streets of Islington. Growing ever more agitated, she stopped sleeping altogether and sat up every night, cradling tiny Margaret and peering through the curtains. Halley realised he had no choice but to move them elsewhere. He chose an old townhouse that put the mass of London between them and the army.

Cripplegate was huddled into the lee of the city wall to the East of London, but retained enough open space to feel like a village.

'You'll feel safer here,' Halley had said on the day they arrived with their wagonloads of furniture and chattels.

Mary had nodded, but it was a small, anxious gesture.

The rooms were smaller than those at Islington, yet more in number, giving the house a cosier feel. Mrs Fletcher and William, the boot-boy, had moved with them, and gradually Mary had settled. Eventually she was persuaded to put Margaret's cot into the nursery, adjacent to the master bedroom.

Now one year old, Margaret was a source of delight. Already skilled at taking all the gloves out of the box in the hall, she now practised the art of putting them all back in. She made her desires known with urgent gestures and a flashing smile, especially when Halley brought home handfuls of Chinese oranges from Covent Garden.

'You're too eager to please her,' Mary would say in tender admonishment as he lay on the floor, feeding his daughter miniature segments.

'How can I not be when she looks so like you?'

'She doesn't.'

'Besides, she must develop her palate for expensive food if we are to find her a suitable husband.'

Mary would reach around him from behind, and in those moments they both knew it was only a matter of time before they added a second child to their family, despite the uncertainty of knowing into which religion they would be forced to baptise it.

The table was laid for dinner. Halley was carving a shoulder of lamb. Through the window, the afternoon light was already fading. In the corners of the panes were the first splinters of frost.

'Monsieur Papin says things are getting worse in France,' he said. 'He expects Protestantism to be made illegal by the year's end, and fears he may never see his home again.'

'Where would we go if that happened here?' asked Mary, her voice tight. 'Holland? I'm not sure I could live in a Low Country.'

'And you won't have to. We're safe here.' Changing the subject, Halley busied himself with the joint of meat and said, 'I'm to report on Mr Newton's revised paper to the Fellowship this week.'

'Is that what's been consuming your time?' she said gently.

'It's a masterpiece, Mary. I just hope I can persuade the other Fellows; they seem unable to drag themselves to Gresham these days. The last Council meeting had to be postponed again for lack of attendance, so we still have no agreement to actually publish the work. Without it, I fear Mr Newton may yet fail to complete the book. He's sent me the first two sections but is still working on the third – and from what I hear, just getting this far has almost cost him his sanity. He collapsed with the effort of it.'

'Perhaps they're too concerned with the state of the country to think very much about philosophy. Now, are you going to serve? I'm hungry.'

'We cannot stop our curiosity because of what may or may not happen with the King,' said Halley, laying a slice of meat on his wife's plate. 'But perhaps you're right. Anyway, we'll find out on Wednesday at the next meeting.'

The Fellows listened as Halley described Newton's work. He looked from one to another hoping for some flicker of interest. Hooke was studiously indifferent; Pepys slouched in the official chair with his fleshy head resting on his upturned palm; Papin was preoccupied.

'Let me reiterate that this is a most remarkable treatise, advancing a number of propositions taken from observations of the world and backed by experiment. On the shoulders of these propositions is built an understanding of the celestial and Earthly motions. Mr Newton concludes that motion is the result of forces acting upon objects. Without a force, an object cannot be made to start moving or change its existing motion. The heavier the object, and by that I mean the more mass it contains, the more force it requires to achieve the same acceleration.'

The audience stared back like sculptures. He raised his voice. 'Chief among Mr Newton's conclusions is the demonstration that heavenly motions are the exclusive results of a force, here called gravity, that decreases with distance in an inverse square fashion. The planets are held to their paths by the gravity emanating from the Sun. Each of the planets subsequently produces its own gravity, which holds its moons.' Still there was no reaction, though perhaps there was a curious nod or two – either that, or the Fellows were becoming sleepy. 'Perhaps once the third section of the book is delivered, this will become more obvious. Mr Newton promises a full discussion of the philosophy rather than the mathematical foundations. Gentlemen, we cannot allow this work to languish. I propose a vote.'

'You cannot do that,' said Hooke, stirred from his reverie. 'It's not for the Fellowship to tell the Council what to do.'

'If the Council cannot drag itself to meet, then the Fellowship must show them the way.' The words were out of his mouth before he knew it. There were at least three Council members sitting around him – Hooke, Pepys and Wren. There was shuffling and muttering among the Fellows, and things threatened to become uncomfortable.

'Gentlemen, with the greatest of respect, all those in favour of publication, please raise your hand,' said Halley.

After an awkward second or two a few arms sprouted, followed by a few more. Halley nodded as one by one the

Fellows raised their hands until it was clear that most of them agreed.

'Very well, it is resolved that the Society wishes to publish Mr Newton's work.' Halley turned to Hooke. 'Let that be communicated to the Council, with the greatest of respect.'

'Waste of time and money,' muttered Hooke.

A larger than usual number of Fellows had trooped through the alleyways to the coffee-shop after Pepys had closed the meeting, but it was not, as Halley had hoped, to discuss gravity.

'Most of the clergy are refusing to read it to their congregations,' Papin was saying, his sharp features drawn.

Pepys blew out his cheeks, cradling a dish of pungent coffee. 'You make too much of this. All the Declaration of Indulgence allows is the freedom to practise your chosen religion, Roman or Anglican.'

Papin's nostrils flared. 'An Englishman should be able to interpret the subtlety better than a Frenchman. James does not confirm Anglicanism as a legitimate faith, guaranteed a place in English society. Let this pass, and the Church of England will find itself outlawed in a matter of years. That's why the vicars are defying the King's order to read it out.'

Pepys shook his head and shouldered his way into the busy crowd.

'I am coming to think that the only safe place is Germany,' Papin said, 'but if England falls, even Germany will be isolated. What is the solution?'

Halley had no answer. The King's Protestant daughter, Mary, now lived in exile in Holland. The previous summer, as the Halleys had been carting their possessions to Cripplegate, the disastrous invasion by Charles II's illegitimate firstborn, the Duke of Monmouth, had culminated in his botched execution, at which it had taken seven swings of the executioner's axe to sever his head. The assizes had then cut a bloody swathe through England's Protestant activists.

Pepys's voice interrupted their thoughts. He had returned trailing Hooke behind him. 'Mr Halley, it seems that Mr Hooke here has something to say on the subject of Mr Newton's ingenious theory. It appears that it was Mr Hooke himself who struck first upon the idea of the inverse square law.'

Other Fellows now turned, their interest piqued.

Halley forced down his impatience. 'Robert, we all suspected the nature of the force but couldn't prove it. How can you claim any part of that?'

'I make no claim on his curves and diagrams, but just as surely I gave him the notion of the inverse square in my letters of 1679. He cannot deny that and neither can you. I told him that orbital motion was a compound of two motions: a direct motion and a deflection towards a centre of attraction. Those are the principles upon which his work rests, and he had them from me – and me alone. Mr Newton is nothing but a plagiarist.'

The Fellows turned as one to Halley.

'Very well,' he said, uncomfortable under their combined gaze. 'I will write to Mr Newton to clarify the matter.'

'You're early,' Grace said, placing the brass candle-snuffer back on the mantelpiece.

'I didn't feel like staying.' Hooke looked around the room disconsolately.

'I've tidied the ledgers in your study,' she said.

'They don't believe me, you know,' he said.

Grace tilted her head. The gesture conjured in Hooke a memory of his mother and the way she would comfort him or his brother. He recounted the conversation in the coffee-shop.

'Forget them, Uncle. Don't upset yourself. I'm worried about you.'

'Why is it so cold in here? There's no fire,' he said.

'Seasoned logs are running short. Any that are left are so expensive.'

'Come.' He took her by the hand, knowing from her warmth that he must feel like ice, and led her towards his bedroom. Her body stiffened and he turned to look at her soft face. 'That's not what I want.' For reasons he still did not understand, all desire had left him. It had vanished so completely it was as if it had never existed at all, but he pulled her onwards.

'Uncle?'

He led her to an oak trunk in the corner of the room. It had not been polished in years and it was scarred with woodworm pores. He removed the old lathe stool that stood on it and lifted the lid.

Grace let out an exclamation of surprise at the contents.

Wad after wad of banknotes, sprinkled with coins.

'This is what I got out of the Fire. We were going to rebuild London with boulevards and squares to rival Paris, but the whole thing got bogged down in legal argument about who owned what. So we abandoned the new plan and just rebuilt along the old streets. They still paid me, of course, and I kept it all here. You know I don't trust banks and investments.'

'There's enough for a lifetime,' she breathed.

'Two lifetimes.'

'Two?' she searched his face.

'What remains of mine, and then . . .' He paused; it felt as if he were about to reveal a great intimacy. ' . . . and then the rest of your life. This is all for you.'

Hooke felt his face flush. Grace covered her mouth with her hand and tears gathered in her eyes. He reached out to embrace her and realised what an alien gesture it was; somehow she had always been the one to embrace him.

'Don't cry,' he said.

'But I don't deserve this.'

'I love you,' he whispered. 'I'm feeling my age, despite what you dress me in, and an old man craves simple pleasures: companionship, and the knowledge that you will be safe when I am gone.'

Grace sobbed on his shoulder. Although the air around them was as cold as before, Hooke no longer felt the chill.

Pepys stood shivering, hunched inside an overcoat on Halley's doorstep. 'Do I catch you at an inconvenient time?'

'No, no. Please come in.'

Ensconced in the drawing room, Pepys accepted a drink from Mrs Fletcher before speaking. 'The Council met yesterday to discuss the printing of Mr Newton's book. We agree with you that it's worthy of publication and that it should carry the Society's imprimatur. And all were resolved that you should be placed in charge of ushering it into print–'

'Of course, I would be honoured.'

'– and part of that duty would involve taking care of the financial side of publication.' Pepys glanced sideways at him.

'You mean I am to be personally responsible for the printing?'

'The Council recognised your interest in the work and thought that you should benefit from its eventual rewards as well.'

Pepys was a terrible liar; he was all but squirming.

'But the Society has published books before . . .'

'The Society has somewhat strained its finances. Our last publication did not do so well as we had hoped.'

'The History of Fishes' by Francis Willoughby, thought Halley, *handsomely illustrated, beautifully printed and totally unreadable*. It had floundered like the creatures it sought so hard to depict.

Pepys continued. 'You must be aware that paper costs are higher than ever; the number of illustrations that Mr Newton requires – each requires engraving – and then there is the editing . . . So much work and money that we cannot take the risk. And who will read this new work? We only have your word that it is *incomparable*. You claim that you believe in it, whereas most of us cannot get past the first page. If you believe in it so wholeheartedly, you should feel it your duty to publish in the Society's name.'

Halley realised he had been backed into a corner. 'Very well,' he said slowly, 'I will publish it; this book is more than incomparable. It is *divine*. I believe it will change our world for ever.'

As Pepys departed, so did most of Halley's bravado. He returned to the drawing-room, feeling foolish under Mary's gaze.

'What can I do?'

'How much will the publication cost?' Her voice was neutral.

'A few hundred pounds at most, I think. I'll make inquiries in the morning. It will stretch our budget . . .' He braced himself for an angry outburst.

Mary shook her head at him. 'Oh, don't be so dramatic. We're far from destitute. We'll make a few economies and all will be well.' She stood and combed her fingers through his hair. 'But I'm afraid the Chinese oranges will have to stop.'

24
Cambridge

'We're done for!' announced the squat little man, chopping the air with his hands. 'The University will be ruined piece by piece.'

If there were two things that hovered near the top of Newton's list of intolerances, they were prevarication and whining. As fate would have it, those were the exact two characteristics that this man wove together in his vain attempt to assemble a personality. In any other situation Newton would have ignored him, but John Peachell was Cambridge University's Vice-Chancellor.

Newton scanned the letter that Peachell had thrust into his hand. It was a royal proclamation. 'You've objected to this, I take it?'

'A letter was sent to London. We await a reply.' Peachell's skin was covered in angry blotches. 'Our riders tell us that Alban Francis left Whitehall to return to Cambridge this morning by coach. He'll be back tomorrow with the King's reply, the day after at the latest. That's why I summoned you this morning. I need help.'

'James will not back down. He cannot show any weakness. This is the devil's work indeed,' said Newton, looking angrily at the missive. It instructed Peachell to confer the degree of Master of Arts upon the Benedictine monk Alban Francis.

Father Francis had become an increasingly visible figure around Cambridge of late, his black habit whispering along the corridors as he glided from one rendezvous to another.

'We cannot have papists in the university hierarchy,' spluttered Peachell, spraying the contents of a tin mug over the white lace of his shirt.

'Oh, I agree. But calm down, Vice-Chancellor. There's a way out of this.'

'Really? How?' Peachell's eyes were wider than Newton had ever seen them before, pale grey irises in yellowing balls. They peered at him from above a brandy-ruined nose.

'We oppose the King on a point of law: no one can graduate without taking the oath of supremacy. Father Francis cannot possibly do that because it's an oath to uphold the Anglican religion.'

Peachell scowled. 'But we've conferred degrees without the oath before.'

'The Moroccan ambassador's secretary? We both know the two cases are entirely different. The first was an honorary entitlement that the secretary could take home as a token of esteem. Alban Francis – correct me if I'm wrong – lives in Cambridge and intends to contribute to University business, as the degree will entitle him to do. How can he do that without swearing one of our most important oaths? Have faith, Vice-Chancellor, no man can be prosecuted for sticking to the law.' Newton forced himself to look into Peachell's uninspiring eyes. 'Bow to this and it will open the sluices for more Catholic effluent. I would rather see our divine halls burn to the ground than be subverted to Roman will.'

Peachell scratched his lank silver hair, covering his shoulders with white flecks. He bit his bottom lip so hard Newton thought he might draw blood. 'Very well,' said Peachell, 'but you must prepare the defence.'

Newton returned to his rooms to find Humphrey pacing.

'Mr Newton, you have a letter.' He handed over the folded sheet. 'And the Master says to remind you that you are late with your lecture notes.'

Newton slipped a finger into the fold and popped the letter free of its wax.

'But what's the point of handing in your lectures when you have no students to talk to?' Humphrey fretted.

Newton glanced at the sheaf of papers on the table. 'My notes for book three: a grand discussion of the moving world, the elucidation of God's Creation. Copy those and hand them in.'

Humphrey looked forlornly at the paper stack. 'But they're not finished.'

'So?'

Humphrey slunk over to the seat to begin his task.

Newton unfolded the letter; it was from Halley. *There is one more thing that I ought to inform you of, that is Mr Hooke . . .*

Newton snatched the quill from Humphrey's grasp and swatted his hand away. 'Fetch more ink, then find me a courier.'

Late the next day, Peachell summoned Newton again. He stamped into the Vice-Chancellor's office and stationed himself with his back to the bay window.

'The King has summoned me to London. I'm to appear before the court.' The spider's web of veins on Peachell's cheeks looked redder and angrier than ever.

Newton could not keep the impatience out of his voice. 'You must take an honest courage with you, Vice-Chancellor. Yours is a just cause, but if you let them, the papists will exploit you.'

Peachell lifted a decanter. It clattered against his glass. 'The judge is to be George Jeffreys.'

That did arrest Newton's attention. Jeffreys had scythed his way through the West Country, dispensing justice through the mass hangings of those revolutionaries who had landed with Monmouth.

'Damn you, Mr Newton, you have driven me to this. So help me, you will stand with me. You and seven others will be appointed by the Senate, and you will accompany me. Get your defence ready; we all stand together. We're little more than traitors now.'

25
Cripplegate

'Edmond, whatever's the matter? You've gone clean white,' said Mary, tugging her shawl up on her shoulders.

'It's a letter from Newton. He's going to withhold the third part of his book because of the argument with Robert. All I did was ask him to consider acknowledging Robert fully.'

'So just publish what you have. The less you have to print, the less the expense.'

Halley looked at her in disbelief. 'It doesn't work like that. The book is incomplete without the third part. It's where he plans to discuss the consequences of his work. Oh, what a mess . . .'

'Anyway, why must those two always squabble so?' said Mary.

'Because men will be men whether they are philosophers or generals,' Halley said, more sharply than he intended. He softened his tone. 'The older Fellows still talk about it. Years ago, Newton arrived unannounced at the Society and promised to send in a paper he had been working on about the nature of colours. Well, it took four sessions for it to be read out, and at the end of it, Robert disagreed with the conclusion. You see, Robert thought that light was like sound – as they all did – and that meant that colours were created in the light by the medium it travelled through.'

She shook her head. 'What do you mean?'

'In the same way that the pipes in a church organ influence the sound of the notes, the timbre and the cadence, so they believed that when white light passes through glass or air or some other medium – water, say – so the colours are created. I think some still believed that the stars are holes in the sky letting through pure light from heaven.'

'So that's why the stars are white,' said William from the door, intrigued by the conversation.

'No, young man, that's mediaeval thinking. The stars are different colours if you really look at them. Arcturus is orange, Rigel's blue; there are colours aplenty if people wait long enough for their eyes to adjust to the dark. Newton argued exactly the opposite to the traditional view. He said that, based on his investigations, white light was the combination of colours. At first, no one believed him. They argued that if you mixed the colours from an artist's palette, instead of white you produced a dark murky mass, but Newton was cleverer than that: he proposed an experiment to prove his hypothesis.'

William took a step forward. Halley alternated his attention between Mary and the boy.

'No one had really done that before. They experimented and observed but then argued about the interpretation. Newton formed his hypothesis from preliminary observations and then devised further experiments to test whether his ideas were true. Do you see, he used his understanding of nature to foretell a future event, to predict what would happen when someone carried out his proposed experiment. He did the very thing that all these fortune-tellers and mountebanks claim to be able to do but cannot: he predicted the future. He said that without such a final test – the crucial experiment – natural philosophers could never be certain that they had understood nature. They'd wallow for ever in hypotheses and never be able to elevate them to theorems. So, to test his hypothesis of colours, he took a prism and used it to turn sunlight into a rainbow of colours. Then he placed a second prism in that rainbow but blocked all but the blue light from entering. He reasoned that if the glass were creating the colours, this second prism would turn the blue light back into all the colours again. But you know what happened?'

William shook his head, his eyes wide.

'It didn't. The only colour that emerged from the second prism was blue, proving that the glass doesn't create the colours, it just separates out what's already there.' Halley turned back to Mary. 'Well, Robert wouldn't have any of it. I think he understood what Newton had done but felt foolish for not thinking of it himself. So, he chose to ignore the evidence and argue against it, and Newton has never forgiven him. He even decided not to publish the work. I cannot allow Newton to suppress more work because of Robert. How am I going to resolve this?'

'I know what will cheer you up,' she announced brightly. 'We'll go to the Frost Fair tonight. It may be our last chance; the crocuses will be up soon.'

He smiled to indulge her, forcing himself to say, 'I shall look forward to it.'

But first he had some serious diplomacy to contrive.

Soft moans of despair escaped Peachell as the delegation arrived at the London courthouse. 'It's unfathomable,' he whined. 'Jeffreys himself should be behind us, he's an Anglican.'

'He knows only how to keep himself in favour,' hissed one of the delegates.

'He's a hanging judge and no mistake,' said another.

Newton had investigated the details of Jeffreys's record in preparation for today, and still fought the nausea his research had conjured. The horror had begun in Winchester, where a Lady Alice Lyle had unwittingly helped two fugitives from Monmouth's routed army. Jeffreys sentenced her to be burned. Only at the King's command was the sentence delayed and commuted to mere beheading, which took place before a silent, disbelieving crowd in the marketplace.

It set the pattern for the justice that was to come. By late September, hundreds of Monmouth's ragged army had been put to their deaths and their quartered remains distributed to the local villages to give them a taste of retribution.

Jeffreys had presided over this squalid mockery of the apocalypse, dispensing judgment to imbeciles and innocents. All of it was designed to terrify and belittle the population and gorge King James's power-lust. If Satan himself had been in control of England, Newton could not have felt more driven to resist.

The Cambridge delegation was escorted to an upper floor of the courthouse, where the noises of the city were unable to penetrate. It was as if the essential march of time had halted and London life would resume only after the judgment was made.

They were led to a wooden stall that resembled a cattle pen. It faced an impressive table, heavy oak with spiral-turned legs and raised upon a dais. It was empty for the moment.

'No chairs?' murmured the Vice-Chancellor, flummoxed. Newton manoeuvred himself next to the quivering man, hoping that he could feed him lines during the examination and stop him swigging from his small leather hip-flask.

The sound of a door opening cracked around the chamber as if someone had fired a musket. Judge George Jeffreys appeared. Although his wig was grey, his face was angular and lean, almost youthful. He radiated confidence rather than the malice Newton had been expecting.

He was trailed by others, all in scarlet gowns except Alban Francis. Garbed in his Benedictine habit, he was chattering earnestly to Jeffreys as they climbed the steps to the central table. The pale light from a window nearby flooded across the polished wood.

'Let him win and we lose the university to the papists for ever,' Newton whispered to Peachell.

Jeffreys spoke up, his voice as stiff as unworked leather. 'Which one of you is Peachell?'

The Vice-Chancellor lifted a tremulous hand. A good six inches taller, Newton felt conspicuous behind him.

'Worthy sir,' began Peachell, 'it gives me extraordinary distress to know that I have offended the royal personage–'

'I don't believe I asked you to speak.' Jeffreys looked at each of the delegation in turn. Newton met his gaze full-on, over the top of Peachell's flaking scalp.

'You, sir, stand to one side so that I can see you.'

Newton moved.

'Tell me your name.'

Newton spoke it clearly, so they all could hear.

'Do you cower in the shadow of your Vice-Chancellor?'

Newton said nothing.

Jeffreys let his gaze linger, some calculation going on behind those unblinking eyes, before turning back to Peachell. 'Do you know why you're here today?'

'Sir, I do not.'

Newton had instructed him on this during the carriage journey: *admit nothing.*

'No? How could you be in doubt of this outcome after your rank disobedience?'

Peachell looked as if judgment had already been passed.

Jeffreys continued. 'Very well. Perhaps you can explain to me why you ignored a direct request, however mildly made, by His Majesty?'

Mildly? thought Newton. *The last letter had stated that they ignored the request at their peril.*

Peachell was stumbling over his words. The court officials were looking from one to another, a mixture of bemusement and boredom on their faces. Alban Francis had a growing look of satisfaction on his face. Newton stepped to the front of the dock. 'The law states that each of us must take an oath before he is appointed to a fellowship or mastership. That oath is to the Anglican Church and therefore impossible for a Catholic to take.'

'Do I take it by this outburst that you fancy yourself as Vice-Chancellor?' Jeffreys paused for his rebuke to have the maximum effect.

Newton saw his chance. 'The King is not above the law. And the law is clear in this matter.'

Ignoring the mutterings that had erupted around him, Jeffreys flipped through a ledger. 'Mr Newton,' he said, studying one sheet, 'can you tell the court what you were doing in London during June 1672?'

Newton's confidence ebbed. 'I attended the Royal Society for the first time.'

'Is that all?'

'And I visited Whitehall to seek an audience with King Charles.'

Jeffreys' eyes were on him now. 'Why?'

Newton had underestimated him. Steeling himself, he spoke up. 'I received a special dispensation.'

'A special dispensation. Was it a special dispensation that allowed you not to take the oath you have just advocated?'

All heads turned towards Newton.

'Yes.'

'So the King is not above the law, but you are.'

Stand firm, Newton willed himself. *No one knows the real reason, and no one must ever know.*

'My dispensation against the oath was a private matter,' he said.

'A private matter?'

'Yes.' They would have to torture him before he would reveal it.

Jeffreys's eyes narrowed and Newton held his breath. If the judge had learned the reason, now would be the time to use it. The pair locked eyes. After an eternity, Jeffreys said, 'Step back, Mr Newton. I'll hear no more from you today. There is no reason why Alban Francis should not be conferred his degree. Mr Peachell, you are guilty of an act of great disobedience. You are to leave the office of Vice-Chancellor at once. You will quit your position as Master of Magdalene and your salary. You may go on your way as a common man again.'

Peachell slumped against the wooden railing.

'As for the rest of you,' said Jeffreys, looking from one to another, 'those children unfit for manual labour are best sent to schools. For all your scholarly endeavours, sirs, you are but children. Fit only for the colleges and hopelessly adrift in the world of true men. Further depositions are neither solicited nor required. I have reached my judgment.' He glared at the men.

Those around Newton dropped their gaze as if in Church, but Newton met Jeffreys's eyes.

'Gentlemen,' Jeffreys said, ridding the word of any respect, 'your best course of action will be to deploy a ready obedience to His Majesty's command in the future. You have set an ill example for those students in your trust. You stand here before me today because of your supposed religion. Therefore let me bid you adieu with what the scriptures say.' His voice rumbled with menace. 'Go on your way, and sin no more, lest a worse thing come unto you.'

Newton contemplated Jeffreys's position – an Anglican doing Catholic work. The judge was destined to burn for ever. The image of it appeared bright and clear in Newton's mind and it made him smile.

The Frost Fair was illuminated by the silver of the Moon and the gold of burning torches. The makeshift wooden shacks, stretched across the frozen Thames, echoed the houses that spanned the nearby London Bridge. The stalls were draped in fabrics and triangular pennants flew from spindly masts.

In places the stalls were squeezed together as tightly as the Arabian souks that the old sea captains described, where you could buy everything from intoxicating powders to women. Here in London the wares were chestnuts and pies, mulled wines and spiced beers: everything necessary to keep the chill at bay, or at least to take your mind off it.

Halley bought two mugs of steaming ale; its heady smell alone was warming. It was an unusual favourite of Mary's, who, despite her passion for the aromatic drink, could only ever manage half the mug before she became giggly and passed the rest to him to finish, a duty he was happy to perform.

It was still quite early and there were several families on the river. One youngster in particular caught Halley's eye. He was a gentleman in miniature, wearing a tightly-fitted coat and hat. He and his father were watching a group of urchins sliding themselves along the ice underneath the wharfs. Halley could see the boy's longing to join in.

'Do you think we shall have a son?' he asked.

'Are you craving an heir, husband?'

Her question evoked a twinge of embarrassment. 'I cannot wait to bring Margaret down here. She'd love all the colours.'

Behind them, a group of men arrived at the tented bar. Stocky and boisterous, they talked with Irish accents.

Mary gripped his arm. 'Let's move on.'

He led her further on to the ice, where they paused as a coach and four crunched past. They watched dogs chasing each other over the slippery ground, as intrigued by the novelty of the Frost Fair as the humans.

Some wherries had been suspended from triangular frameworks to create swings, and Halley urged Mary towards one where a couple were disembarking. A familiar figure with a silver frizz of hair was watching the swinging boats as if hypnotised.

'Mr Newton?'

The man turned, his face impassive. 'Mr Halley, a pleasant surprise.'

'I had no idea you were planning a London visit.'

'I return tomorrow.'

'Allow me to introduce my wife, Mary.'

'Mrs Halley,' Newton nodded.

'I have been writing to you today, sir, concerning your latest letter,' said Halley. 'This business with Hooke, sir. It is a trifling matter when held against your achievements in philosophy.'

'Philosophy.' Newton said the word as if it were a profanity. 'A man might just as well embroil himself in lawsuits. The trouble with Mr Hooke is that he cannot bear to see a procession going along without him.'

'Sir, you must publish the third book. Why sow, tend and nurture your fruits, only then to walk away at harvest time? You have so much to teach us.'

'Mr Halley, I have given you some trouble, and I somewhat regret my hasty letter to you, though I stand by its sentiments. I desire a good understanding to be kept between us, and have decided on my course of action. I will not suppress part three, but I will revise it. Are you familiar with the story of Galileo's book, *Discorsi*?'

'I have read it, of course. His mathematical summary of Earthly motion.'

'I mean the composition, the way it came to be written in such mathematical tones.'

Halley tried to remember his conversation with Viviani in Rome, but eventually shook his head.

'His previous book, *Dialogues,* was written as an accessible discussion that could be read by everyone, including the cardinals of the Inquisition. That was Galileo's mistake. So, he couched the *Discorsi* in only maths, measurements and experiments. The cardinals couldn't understand a word of it.

Had they been versed in geometry, they would have seen that Galileo wrote the truth. It was a masterpiece of composition, a work of such importance that only those intelligent enough to understand it could read it. I will write book three in the language of mathematics, so that only other mathematicians can understand it. It'll prove to everyone that this is not the work of some little . . . smatterer.'

Relief blossomed in Halley. 'However you care to deliver your insights, I and the rest of the world will be most pleased to receive them – be in no doubt about that. I have already engaged a second printer; we can have it published in no time. No time at all. Sir, you do not need me to tell you that you have seen closer than any of us into the mind of the divine Creator.'

A faint warmth lit Newton's face. 'Then I bid you goodnight; I'm on the early coach back tomorrow. Fear not, Mr Halley, I will send you the completed work soon.'

They parted, and Halley looked over at Mary.

'You see, all resolved,' she said.

He smiled at her little red nose. 'Let's take that swing.'

26
London
1687

King James II should have been handsome. He sported the cleft chin and dark eyes of the Stuarts, but somehow failed to wear them with the same aplomb as his brother. Perhaps it was the darting of his narrow eyes, or the creases in his brow; whatever it was, he looked like a poor copy.

He was perched on a receiving throne, leaning heavily on one arm and turning the pages of the book listlessly with the other, and his auburn periwig had such tight curls it could have been a doormat for wiping boots.

'A handsome volume,' said the King. Around him, his advisers preened themselves and looked uninterested.

'It is an incomparable treatise by an English scholar, Your Majesty,' explained Halley.

The King tipped the book and studied the gilt lettering on its spine. *Philosophiæ Naturalis Principia Mathematica.* 'Newton.' His face tightened and he looked up. 'Would that be Newton of Cambridge? The same Newton who opposed me and Alban Francis?'

Halley was prepared. 'I know of your fondness for the naval arts, indeed your aplomb in command of a vessel. I think the book's value to English shipping is beyond measure, sir. It allows us to predict the height and times of the tides with extraordinary precision. Decades ago, the German astronomer Kepler drew attention to the fact that the tides are somehow linked to the movement of the Moon. Mr Newton has now described in precise mathematical detail that this is because the Moon releases gravity, a force that attracts our oceans, pulling them into tides, whereas the solid rocks of land can resist. Mr Newton goes on to explain the different heights of the spring and neap tides.'

James' face had softened. Halley took it as a cue to continue.

'The Sun also releases gravity to keep the planets in their courses, and this too raises tides on Earth. When the Sun and Moon are aligned with Earth, as they are at new moon and full moon, the gravitational attractions of the Moon and Sun add up, and create the higher spring tides. When they're at right angles to one another, they oppose, and the oceans are pulled in lower, or neap, tides. We can calculate with accuracy the times and the heights of future tides, and plot when is safest for our navy to sail – or when it may be advantageous to draw the enemy towards us on to the sandbars.'

'Remarkable, and strategically important too.' James looked around at his advisers. 'Do you not think so, gentlemen?'

There was some muttering and half-hearted nodding.

'I miss the sea,' continued the King. 'For all its uncertainty, you knew where you stood. None of this lot can understand. Are you run aground these days, too, Mr Halley?'

'I am.'

'Wife and family?'

Halley nodded.

'Be grateful you don't have a country to run as well.'

'Yes, sir. One daughter and another child on the way are enough for me. I hope for a son.'

'Don't we all, Mr Halley?'

'Your Majesty?'

There were some sharp looks from the advisers.

'Haven't you heard yet? I thought the mob would be gossiping about it by now, I've gone beyond expecting them to celebrate it. I'm to be a father next year.'

'My congratulations, sir.'

A smile attempted to find purchase on James's face but served only to make him look unhappier. 'Do not let me detain you any longer, Mr Halley. We thank you for bringing word of this new knowledge today and for the book. Mr Newton, it seems, has gone some way towards redeeming himself.'

Halley bowed deeply, backing away as he did so. He was on the verge of turning away when James called his name. Half question, half statement; it was an exclamation of recognition.

The astronomer froze.

'Yes, of course: Halley. Didn't I meet your father once?'

Halley looked the King straight in the face. 'If you did, sir, he took that honour to his grave.'

James watched Halley until the tall white doors closed.

'Don't match wits with the King,' hissed a familiar voice in Halley's ear.

The astronomer jumped round to see Winslow's pallid face staring at him. 'You?'

'Who else? Just because the monarch changes doesn't mean the rest of us have to. I'll show you out.'

Halley followed the spymaster through an ante-room where the day's hopefuls were milling about, waiting for an audience.

'Tell me about Newton,' said Winslow casually.

'Nothing to tell, except that he's a man of extraordinary mathematical ability.'

Winslow pulled a sour face. 'I don't trust him. Isn't it time you went on another of your voyages?'

'I'm content to be at home with my wife and family.'

'That's not what you intimated to the King, and we need intelligence more than ever. Not that rubbish about feet and inches you peddled us last time; proper information about army sizes and the mood of the people. If the King does have a son it will establish a Catholic succession in England. We need to know who will continue to support him and who will turn. The Dutch certainly won't be pleased.'

They reached the palace entrance hall.

'The affairs of state are far from my concern,' said Halley.

Winslow tutted. 'They shouldn't be. It's your children who will have to live with the consequences.'

Halley stood in the gloomy entrance hall at Gresham and knocked again on Hooke's door. When there was no answer, he pressed tentatively and found it unlocked. Sticking his head inside, he called for both Robert and Grace, but the apartment was still.

He thought about waiting, but he couldn't risk Grace coming back first.

That was the last thing he wanted. When he chanced upon her at Society meetings, she would look through him as an optician might a useless lens. Yet he could still taste her kiss. The memory would assault him when he was least expecting it; worst of all was when he lay with Mary. *Mary!* Imagine her anguish if she ever found out.

He considered leaving the copy of *Principia* but decided against it. Hooke's demands for recognition had made Newton weaken its few small references to the Gresham Professor. On one proof sheet, the author had struck out 'the very distinguished' to leave 'Hooke'.

Halley had not dared replace the deleted words. The only thing he had been able to do was move his own name to after Hooke's in the part where Newton described the origin of the inverse square law. He should at least let Hooke know what to expect.

He looked once more around the apartment. All the furniture had been pushed to the sides to clear the floor. He called again and was turning to leave when a weak voice called out. 'Edmond! Is that you? Edmond, in the name of God, help me! Help me, please.'

Hooke appeared in the doorway, his face drawn.

'Robert, are you in pain?'

'It's Grace,' Hooke said, and disappeared. Halley raced across the room, dropping the *Principia* on the table as he went. At the top of the twisted staircase that led to Grace's chamber he found Hooke hunched over her bed.

No!

It couldn't be Grace lying there, an immobile sack of skin and bones. Her face was white and slack; her sky-blue dress had been loosened, but she still had on her tiny shoes.

Halley dropped his hand to her foot, touching the leather. 'Oh, Grace . . .'

Hooke shot him a look. 'There must be something we can do.' He gripped Grace by the shoulders, his fingertips and cracked nails pressing into what looked like waxed parchment.

She looked impossibly small. Whatever had made her unique had already flown. Halley looked to the ceiling. *Where have you gone, Grace? I cannot believe you've just disappeared.*

His breathing grew ragged and his eyes felt tight. He fought for control and reached over Hooke's bent shoulders to remove the man's arms from her body. He spoke as tenderly as he had ever done to Mary. 'Come on, Robert. There's nothing more we can do now.'

He stole a final look at the figure on the bed and guided Hooke to an armchair downstairs. 'Where do you keep your cognac?'

Hooke pointed to a cabinet where Halley found glasses and a dusty bottle. He poured two drinks, downed one and handed the other to Hooke.

'Best take it down in one.'

Hooke stared morosely at the fluid. 'Will you find Kit? Someone must check the house and the body.'

'No one will suspect the plague. London's been free of it for years now.'

'She was only twenty-seven. Why would she die?' Hooke looked accusingly at Halley. 'She'd been out to her dance lesson and come home determined to teach me. We cleared the chairs, but I was hopeless. At first I thought she was breathless because of the exertion – it's usually me who runs out of breath first. Then she said the room was spinning, so I helped her to sit down. Then she said she needed to lie down.'

'When did it happen?'

'Lunchtime. At breakfast she was fine, chattering about the fair on Spittlefields. I helped her to bed and made up a tonic for her, but she couldn't drink it. She said she wanted to sleep it off, but not more than an hour later I heard her scream. She said her joints were aflame . . .' Hooke struggled for control. 'She was worried . . . she was slipping into the pit.'

Halley leapt up to pour more cognac.

'I mopped her brow and talked to her, then left her to sleep, but when I returned, an hour or so later . . . she was gone.'

Halley sighed. 'Tell me what to do.'

'Please fetch Kit. She must be examined, and she would not have wanted it to be you.'

A twinge of shame went through the younger man. 'I wasn't suggesting . . . I'll go and find Kit. Shall I also inform an undertaker on your behalf?'

Hooke wept in silence.

On the day of the funeral, Halley and Mary arrived at Gresham to find every window, every mirror, every display case and glass-fronted cupboard draped in black cloth. He shared a look with his wife and saw that she was thinking the same as he was. It was an old superstition that reflections must be banished to prevent the deceased's spirit finding a way back to Earth.

Perhaps Hooke had done it more out of tradition than fear. The thought made Halley wonder how he might behave in a similar situation. Mary was about the same age as Grace. The horror of a sudden emptiness at his left shoulder, coldness in their bed, or a silence at mealtimes made him recoil.

The coffin was still open, but he did not want to see the body again, so, when Mary drifted over to pay her respects, he stood by the tall clock where Grace had kissed him. He looked over at the staircase that led to her room. More people arrived, clutching the notification cards adorned with skeletons and gravestones.

Mary rejoined him to take their place in the cortège and they watched solemnly as the coffin lid was secured. They held each other as they followed the black coffin out of the apartment and into the night. When they reached the quadrangle they were handed torches, and the procession snaked from Gresham to the nearby church.

As they filed into the pews, the torchlight turned the mourners into dancing shadows. 'Mary,' Halley whispered, 'you know we have often discussed religious matters, and my opinions have given you some concern.'

She glanced over. 'Is now the right time for this conversation?'

'I want you to know that I do believe.' *At least, I want to believe.* He had to consider that Grace was somewhere; otherwise the weight of the situation, and the knowledge that one day he would lose Mary, would have been too much to bear. 'And I love you more than anything.'

After the service they went out into the moonlit graveyard. Hooke watched as the coffin was lowered into the ground. A new black suit was rucked over his back and his hair was a lank, greasy mass. He had said little all evening, barely acknowledging anyone's presence. Now, as he watched his niece's final journey, he was stooped so low that he looked as if he would roll into the grave beside her.

27
Cripplegate

'I'm not sure we should have gone ahead with this,' Mary whispered.

'Nonsense,' said Halley. 'These plans were in place long before Grace died.'

'But it looks disrespectful.'

'I asked Robert if we should cancel and he said we were to enjoy ourselves.'

They stood in their hallway welcoming a steady procession of party guests. Carriages and horses milled outside, unloading more and more people; others arrived on foot and struck up conversations of their own as they waited in line to be greeted.

Mary was wearing a dress of pale yellow. Her forearms and collarbones were on prominent display, lending her an air of casual disarray. When Halley had remarked on her appearance before anyone had arrived, she had been quick to say, 'It's the new style, I'm told. Don't you like it?'

'Quite the opposite,' he had said, and kissed her, squeezing her a little more tightly than might have been advisable in her condition.

Now she charmed the guests with her effortless conversation while Halley welcomed them in more robust fashion, gripping male friends by the wrist and clapping some on the shoulder, but they fell silent at the next guest.

'Robert,' Halley said.

Hooke had a faraway look.

'It was brave of you to come out tonight, Robert. I'm honoured to have you here.'

'Nonsense. Life goes on, eh?' Hooke patted Halley on the arm, still not meeting his gaze. 'But I'm sorry she's missing it. She did love social engagements.' He spoke as if she had been called away on a family visit and would return in a day or two. 'Now, where's the wine? I've had a hectic day.'

'Robert, I should warn you that Mr Newton is here.'

Hooke frowned. 'Is there anyone else you should have warned me about?'

'No, it's just . . .'

'I found the copy of *Principia* you left for me.'

'I'm sorry. I had intended to present it to you, but with the situation . . . Did you manage to look at it?'

Hooke nodded. 'Enough to see that it is of his normal – lamentable – standard, but at least there's some small acknowledgement of our contributions.'

Halley released the breath he had been holding. 'Let's find you a drink. Go on through to the reception room. There should be some cordial to cheer up the Bordeaux. Mrs Fletcher was boiling apricots this afternoon – always a good sign.' She had been making lemon syllabub as well, although Halley was secretly hoping that not everyone would find it, ensuring some leftovers for tomorrow.

Halley's heart turned over at the charade of cheerfulness. If anything, the Gresham professor looked worse than he had at the funeral. He was still wearing his black mourning suit, and his eyes were even more baggy and ringed. Halley exchanged a look with Mary.

'What can we do for him?' she asked.

'Keep him busy, I suppose.' Halley shrugged and they turned back to the procession of guests.

When the stream of new arrivals petered out the couple entered the party themselves, swiftly separated and occasionally pressed back together. As Halley mingled, treading carefully to avoid the ladies' hems, he thought about the larger rooms at Islington.

'Edmond, I know this is not the most appropriate occasion but I have a proposition for you.'

He looked up to find Pepys standing in his way. 'Go on.'

'I think when you hear me out you'll agree that it's best to conduct this in private. I promise not to keep you from your guests longer than is strictly necessary.'

Halley scanned Pepys's chubby face, then led him to the study, leaving the chatter and laughter behind. It was warm inside the small room, with an aroma of leather books.

Pepys drained his wine flute. 'I have a ship in need of a captain.'

'Go to sea? Impossible. You know I'm to be a father again. Have you been talking to Winslow?'

'Who? It's only the Thames estuary. Surveying. There's good money in it. Though you scarce seem to need it.' He lifted his glass. A shaft of sunlight caught it in such a way that the black hue of lead was visible where the glass thickened at the bulbous rim.

Halley decided not to explain that the glasses had been a gift from Mary's father. 'There are plenty of good sea captains wanting to go back on the waves. One finds them in every tavern.'

'We need people we can trust.' Pepys took a step forward. 'There are signs that we might need to defend the approach to London.'

'From the French?'

'The Dutch. James has had a son, and his daughter Mary has lost her claim to the throne. The intelligencers tell us that her husband was counting on being crowned King of England.'

'Ah, you *have* been talking to Winslow.'

'I don't know who this Winslow fellow is. I'm talking about William of Orange. We think he is preparing to invade.'

Halley exhaled.

Pepys continued. 'Rochester and others may be stoking William's passion. With him in power, they think they can restore Parliament. A crisis is coming, Edmond. Don't be caught on the wrong side of it.'

'I have nothing to fear.'

'Everyone is under suspicion. The King has been asking about you. He's made some link to your father and the Earl of Essex. A show of loyalty now could prevent any further investigation.'

'There's to be an investigation?'

'Not as things stand, but action may prevent one.'

'This is some news you bring me. Very well, I will join the survey. I will tell Mary in the morning.'

Pepys looked at him approvingly. 'You're making the right decision.'

'I'm trapped by other people's politics. I appear to have no choice.'

When Halley stepped out of the study some moments later, he walked into an unnatural hush. For a moment he thought that everyone had left, but an anxious glance told him that there were plenty of people still present.

A few steps further, and the unmistakeable sound of raised voices reached his ears.

'You are nothing but a common thief!'

'You did nothing but pass off others' work as your own.'

Halley worked a path through the guests in the hallway towards the drawing-room. Some were staring into their glasses; others were craning to catch a glimpse of the combatants.

He saw Mary, looking fraught, and worked his way to her.

'Where have you been?' she whispered.

The voices rose again.

'I don't know what you mean,' said Hooke.

'Then you are unfit to take part in the discussion as you have not read the work of your predecessors. Ask Sir Christopher if you don't believe me. He also had the inverse square before you.'

'There is nothing you can say that will excuse the erroneous diagram you sent me in '79. You clearly didn't understand orbital motion then, did you, Mr Newton?'

'A careless slip with a pen that you mistook for a spiral.'

'Tell the truth!' roared Hooke. 'I corrected you then and I'll correct you again now.'

A circle of onlookers had formed around the two men and Halley squeezed between them. 'Gentlemen,' he said, stepping into the ring.

Newton was leaning towards Hooke, chin jutting and nostrils flaring. Hooke was making vicious stabbing gestures to emphasise his words.

'Gentlemen!'

Newton glanced at Halley and the wildness in his eyes began to subside. 'If you must know, Mr Hooke, your harrying letters gave me no peace at a time of great distress. I was recovering from the death of my mother, and even you can appreciate the impact of a death in the family. There, now you know the full of it.'

Hooke turned to stone, pale lips parted.

Halley seized his opportunity and slid a comradely arm around Hooke's shoulders. 'Come on, Robert. Let's get you home.'

Newton watched, his expression uncertain. Bizarrely, it reminded Halley of the look that crossed little Margaret's face when confronted with something she could not comprehend.

'Excuse us,' said Halley. 'I think Robert may be feeling under the weather.'

Hooke said nothing as Halley led him from the room. Behind them, conversation resumed.

It took most of the next morning to tidy the house. They threw open the windows, admitting the scent of flowers and the loaded smell of pollen. Mrs Fletcher found glasses in all the tiny nooks and crannies where guests invariably choose to hide them. Mary collected up the laundry amid protests from Mrs Fletcher, who insisted that it was her job. William stripped off his jacket and rolled up his sleeves, the way he had seen men do in the Islington fields, and helped Halley to rearrange the furniture. And Margaret watched all this activity, intrigued.

'I do hope Mr Hooke will be all right,' said Mary.

'He was more himself again by the time I helped him into the carriage.' Halley swung an armchair into place.

'I'm glad you insisted on him being driven.'

'He told me he often walks at night now.'

'All the same, I should only have worried more.'

As troubling as the episode with Hooke and Newton had been, Halley's mind was preoccupied with the conversation with Pepys. In the middle of the night he had resolved to forget the whole thing, but by the morning he had come to accept that Pepys's proposal was the only solution. As long as James was alive, the Earl of Essex's ghost would haunt him. He had to show his loyalty somehow, and this could be an opportunity to make some money at the same time. The rent from his father's properties was adequate, but hardly a fortune, especially after he doled out half to Joane.

'Mary, I've been offered a ship.'

'Are you going to take it?'

'I am.'

She stopped folding the lace napkins. 'But I thought you were happy at home.'

'I am.'

'Then why?' She searched his face. Halley almost started to tell Mary everything – the truth about the day his father had turned up at Islington, his suspicions about the old man's subsequent demise, James's new suspicion, and the threat of a Dutch invasion. He wanted to confess it all, to help her understand, but he stopped himself. Only by not telling her could he protect her if things went wrong. These days it seemed knowledge was like a spice from Asia; the very smallest quantity came at the greatest price.

He could see in her eyes that she knew he was concealing something.

'Don't treat me like other men treat their wives. I thought we were different.'

'We are different, except in this one matter. I cannot reveal myself.'

'Is it another woman, now that I am confined?'

'No, Mary, no. You know I love only you. One day, I promise, I will tell you what compels me to do this, but it will not be today. You have to trust me.' He hoped that she would not probe further.

She hugged herself and looked away, blinking.

'Do you trust me as I trust you?'

It took her a moment to nod. He pulled her into his arms. She did not resist, but she did not hug him back either. He kissed the top of her head. 'I'll be back before you know it.'

'Don't play with me, Edmond,' she said. 'I couldn't bear it.'

'I never will, I swear.'

28
London

It took eighteen uncertain months for James's reign to break. When it did, it was as sudden as a bridge collapsing. It was in the early hours of a frosty night illuminated only by starlight. Hooke was out walking, wandering really, when he observed a carriage pulled by galloping horses speeding out of St James's Park.

The King is fleeing!

London was already buzzing with the news that William had landed at Torbay and was making steady progress towards London, taking Plymouth, Salisbury and Hungerford in a coordinated surge. Messengers reached London every day carrying stories of further relentless progress. Only at Reading had the King's Irish soldiers put up any form of resistance. Even then they had been soundly beaten by the Dutch and forced into retreat towards London.

As William's army approached, Hooke had noticed a precipitous rise in the number of men prowling the streets at night. They would accost passers-by and question them about their motives for being out, then claim that they were watching out for Catholic incursions or attempts by the papists to defend the city. Hooke mistrusted all of them and thought many were just spoiling for a fight, especially after he heard that there had been skirmishes to the north of the city.

Hooke saw the King's carriage head west and out of London. He was not the only spectator. A rabble, drawn from the side-streets by the sound, gazed after the receding vehicle almost forlornly, unsure what to do now their greatest wish had been realised.

'Wallsworth, he's a papist,' said one voice, raising a chorus of gruff agreement.

Hooke watched as they battered down a nearby door and stormed inside. He turned to leave, bumping into a man in his nightshirt who had stepped out to see what all the fuss was about.

'Go back inside and barricade your doors,' Hooke told him, before hastening to Gresham.

As the December air bit into Hooke's face, he clutched the wooden balustrades on his observing platform and peered out over the darkened roofs. More and more people were surging into the streets now, spreading as fast as the fire had decades before, and Hooke spent the rest of the night listening to the cries of the mob and those subjected to its rough justice. The first twists of flame began shortly after as the Catholic churches were ransacked and their furniture used as kindling.

Dawn's light brought no warmth but a certain grey calm to the city. Smoke palls drifted, and the streets were filled with people carrying armfuls of panic-bought provisions. As Hooke threaded his way through the crowds, rumours, gossip and the occasional recrimination caught his ears.

Overladen carts lined up at makeshift roadblocks, creating impasse on the streets. Hooke suspected that the self-appointed militia who had taken to the streets last night were now manning those barriers. He shuddered at the thought, especially when he saw a line of men and women kneeling against a wall, their wrists bound with bailers' twine. Men dressed in civilian clothing and carrying bludgeons stood alongside them. Hooke noticed that a few sported wide, white collars that harked back to Puritan garb.

Negotiating Bishopsgate without being challenged, he dipped into the alleyways leading to the coffee-shops and let out an audible sigh of relief when he found Wren in Garraway's.

'We'll soon be in the grip of mob rule,' said an anxious man.

'The Lord Mayor is doing what he can to bridge the gap,' explained Wren. 'At dawn this morning he moved on the Tower, which is now in our hands.'

'What of Jeffreys? I heard they picked him up in a tavern in Wapping. They say he was disguised as a sailor.'

'He's in the Tower for his own protection.'

'Protection?' The man sneered and spat on the floor.

It soon became difficult to follow the threads of conversation. Everyone was speaking at once, stoking each other's fears.

'I'll tell you something for certain,' said a weary old merchant. 'If James has gone, he won't have paid that Irish army of his before taking off, and we all know what an unpaid army does to recoup its losses.'

'I've heard they're marching on London already,' piped up another.

Wren gestured for the men to be quiet. 'Gentlemen, they won't march into the city. It would be madness.'

'They won't march in daylight, that's certain,' said the merchant. 'They'll wait for night and slit our throats while we sleep.'

As darkness laid its blanket over the city, the streets filled up once more. The militia gathered in ominous silence at the city gates, blocking routes into the city with rubbish and anything they could loot from Catholic houses. Hooke once again stood vigil on the observing platform, scanning the horizon from the Tower to the walls of St Paul's and on to Westminster. He listened, alert at any moment to hear the first wail of the Irish onslaught.

He was still there at dawn, frigid and shivering. The night had passed and the attack was stillborn, if it had ever been intended. He learned later that groups of Irishmen were wandering the countryside that night, and rather than slitting throats they were begging for help to get home.

James's last hope had crumbled.

London was William's for the taking.

The cannons of London Tower boomed, rolling their thunder around the city and sending their smoky white billows across the well-wishers on Tower Hill. It was Coronation Day, and the volley marked the beginning of the festivities.

En route to the city that morning, the Halleys had insisted on hauling Hooke out of Gresham, where they had found the man absorbed by a piece of clockwork.

'I thought you were off sailing,' said Hooke, looking bewildered.

Halley laughed. 'I finished that months ago.'

'I haven't seen you at meetings.'

Halley hesitated. There was a pall over the Society these days. Hooke's decreasing vigour and interest had translated into a drastic reduction in the number of experiments. Though it pained Halley to admit it, he was finding it increasingly difficult to motivate himself to attend, and his spell of surveying had rekindled his passion for the water. He was already planning further voyages.

Mary, sensing he was at a loss, said, 'I'm so pleased to see you looking more yourself, Mr Hooke. You had us quite worried for a while there. We came to ask you to accompany us to the Coronation.'

'Is that today?'

They set off and found a space close to the double stone towers of Westminster Abbey.

'We'll see everything from here,' said Mary, bobbing excitedly at Edmond's left arm. They waited in the chill February air, insulated by the mass of bodies around them. The happy morass of people not only filled the streets, they hung from windows and clung to chimneys, all hoping for a glimpse of royalty.

Mrs Fletcher had stayed in Cripplegate with Margaret and baby Katherine, who had been born soon after Halley had returned from the Thames estuary. He had been so caught up in the joy of a second safe delivery that it had only occurred to him some hours after the birth that he had been hoping for a son.

The first sign of the serpentine procession was the women scattering flowers and herbs on to the river of blue velvet laid for the occasion. Behind them a knight in full armour rode a white stallion. Drummers and trumpeters stoked the patriotic fervour of the onlookers, provoking them to cheers. Halley placed his arm around Mary and squeezed her tightly, lifting her on tiptoe so she could see more clearly. Red-robed officials, black-robed Anglican clerics and choristers in angelic white walked past at a stately pace. A pair of men in heraldic tabards followed, each carrying a royal sceptre – one for William and one for Mary. The Lord Mayor followed, his chain of office glinting in the late winter sunshine and his trailing scarlet robe held up by an attendant.

The golden canopy that sheltered the King and Queen had just bobbed into view at the back of the line when Halley heard Hooke say, 'By the grace of heaven!' It was followed by a snort.

Halley followed Hooke's gaze and caught sight of a tall man with a silver frizz of hair sticking out from beneath the wide brim of a black hat. He was walking with his cleft chin held high.

'How did he get there?' asked Hooke.

'He's a powerful man now. His stand against Judge Jeffreys has been recognised. Cambridge has put him forward as its Member of Parliament.'

Newton walked by, not five feet from them, but showed not the slightest hint of recognition.

Hooke watched him pass. 'Powerful for now,' he said, 'but everyone's ignoring his secret: why has he never signed Cambridge's Oath of Allegiance? What's he hiding? I've no more energy to fight him, but someone will. And when they do, Mr Newton will get his comeuppance.'

29
Hampton Court

The wherry bobbed to a stop under the boot of a royal guard on the quayside. Newton held up his invitation. The guard looked offended at having to lean forward in his stiff armour. He made a lingering appraisal of Newton before dropping his eyes to the printed card. Newton felt the smallest twist of uncertainty in his stomach.

The court of William and Mary: I might as well be stepping ashore in a foreign land.

'The Dutch philosopher, Christiaan Huygens, adviser to the King, has invited me here,' he said.

The wherryman leaned on his staff to hold the vessel steady as Newton rose to disembark. The guard removed his foot abruptly and the wherry rocked; to his horror, Newton imagined taking a tumble into the Thames. With a rather inelegant display of arm-flapping, he regained his balance and stepped on to the wooden quay.

A row of soldiers carrying short pikes remained outwardly impassive, but Newton knew they were sneering at him. He shot each one in turn a filthy look as he strode past.

Beyond the guards, set back from the top of the grass embankment, towered the red brick walls of Hampton Court; across its roofline ran a regiment of chimneys. The building was another monument to the profligacy of Cardinal Woolsey and Catholic self-aggrandisement. At least its reclamation by a Protestant monarch was a step in the right direction.

He was assailed by forbidden thoughts. *One step at a time*, he warned himself. *An Anglican England is better than a Catholic one.*

He forced himself to concentrate on the patterns in the brickwork as he climbed the steps of the embankment. Ignored by Charles and James, the palace had been left to crumble over the last two decades. Craftsmen and servants milled about its entrance like bees returning to a hive. Carpenters carried wood, cooks carried bowls and boys dashed to and fro with messages and packages, all caught in the whirlwind of the royal court.

Two men were walking towards him. Instead of workmen's aprons, they were dressed in the clothes of gentlemen. Newton fancied that the one on the left must be Huygens. Stooped, with a considerable paunch, he wore an expensive wig of dark curls and a rust-coloured jacket. He held the garment closed in a rather self-conscious way. The stiffness in his walk betrayed his six decades, but his eyes were still bright and his full lips wore a ready smile.

'Mr Newton, finally to stand face to face!' Huygens spoke with a strange accent, unlike the Dutch Newton had grown accustomed to hearing. He fancied it must be the influence of Huygens's decades in Paris.

'A welcome respite from the volumes of correspondence I am now forced to endure. There seems hardly a day goes by without another letter from somebody previously unknown to me.'

'A tribute to your *Principia*. Copies are already highly sought and rather difficult to come by.'

'Perhaps Mr Halley should have printed more. Still, I'm already planning a second edition to address the problems I left unsolved.'

'A second edition?' Huygens's companion queried in a different, softer accent.

Newton shifted his attention.

'Do excuse me,' said Huygens. 'May I introduce Nicolas Fatio de Duillier, who I dare say is a name to watch. He is a gifted mathematician from Geneva.'

Newton caught the young man's gaze and held it, entranced by such large eyes in an elfin face. 'Delighted to have you in London.'

'The pleasure, sir, is mine alone.' Fatio bowed extravagantly, managing to keep eye contact all the way down and up again.

The exception to Fatio's beauty was his nose, a large, rather ugly thing that dominated his face. At least, that was the impression his bashful expression gave, but Newton rather liked the fact that the young man was not quite perfect.

'Do you now make your home in the Low Countries, Monsieur Fatio?' he asked, the French title sounding awkward from his lips.

'No sir, I teach mathematics in Spittlefields.'

Newton raised an eyebrow. 'A resident of London. Then perhaps we will see more of each other.'

'I hope so. You say that you have left problems unsolved. What, may I ask, are those?'

'The Moon's orbit. Our nearest neighbour is the only member of the celestial family that doesn't obey Kepler's laws. The shape of its orbit appears to change during the month, speeding up and slowing down seemingly at will. I intend to prove that it is being pulled by the gravity of both the Earth and the Sun, and that is what causes it to dance. Once I do that, I will have explained how all celestial motion is due to gravity.'

They made their way into the first of the palace's courtyards, where there was another tide of servants, some hanging out of windows with polishing cloths fluttering across the glass, some carrying furnishings and tapestries. All was noise and activity, and Newton had to raise his voice to be heard. 'Tell me, how is my work being received on the continent?'

'I am becoming your champion,' said Huygens with a smile. 'My latest correspondent is your own countryman,

John Locke. He says the mathematics is beyond him, but he sought a clarification that your basic ideas were sound.'

'You can follow my geometry?'

'I can, sir. You are to be applauded for your system of the world. It explains more than I ever thought possible for a man to know. All Galileo's Earthly motion and all Kepler's celestial motion resolved into a single unified theory of gravity. I have made this clear to Mr Locke.'

'He is in exile, is he not?'

Huygens nodded. 'But I think that with the Stuarts gone, perhaps he would like to return. He would certainly like to meet you.'

'Mr Newton, when I arrived in England, I stepped ashore a Cartesian, but now I have studied your book, I see Descartes as an empty imagination,' said Fatio.

Newton could not help but stare at the young man. 'Tell me, Monsieur, are all the Swiss like you?'

'How do you mean, sir?'

'I mean so . . .' Newton paused, ' . . . delightful.'

'I could not say.'

'Then allow me to say what our modest young friend cannot,' said Huygens. 'He is gifted in the mathematical arts. He is highly regarded by King William, and has already distinguished himself with Cassini in Paris and Leibniz in–'

'Leibniz?'

Fatio continued effortlessly. 'He rebuffed me, sir. When Christiaan here tried to engage the two of us in mathematical correspondence, Herr Leibniz claimed that he saw no need.'

Newton caught the mockery in the title and his lips curled into a smile. 'I see.'

'I think perhaps Leibniz was claiming that he could teach you nothing,' said Huygens.

Newton ignored the Dutchman, giving Fatio his full attention. 'Well, Monsieur Fatio, I was about to say that if all the Swiss are like you, I would be forced to consider it the most charming place in the world. As it is, I am content that the best part of Switzerland has chosen to visit this country.'

The young man flushed scarlet, like a girl in over-laced stays. Newton thought how becoming it made him look, innocence and virtue balanced against intellect and manners. A primal feeling stirred, and Newton thought he might blush as well. At first the tingling shocked him – a distilled version of what he had battled with in his younger years. Thankfully this was not as frightening as the bodily fires that had raged when Wickens laughed, or stripped to wash in the summer.

Newton tore his gaze away and turned to Huygens. 'How long are you in London?'

'All too briefly, I'm sad to say. I return to Holland in a few days, which makes our business here all the more urgent. Mr Newton, have you thought of how the King might reward you for your loyal service?'

'I have. The position of Master at King's College is vacant.'

'It seems a modest move for one such as yourself.'

'Maybe, but I believe it would be a good move for me.'

Huygens nodded. 'I will make inquiries at once. I believe this matter could be brought before the King himself this very afternoon. Gentlemen, please excuse me. I will see you both at dinner.'

Newton smiled at Fatio. Neither seemed quite able to speak. They were standing at the entrance to a passageway leading out of the courtyard.

'It will take us to the back lawn,' said Fatio.

'Shall we?' Newton gestured him on.

They wandered through, forced into closer proximity by the narrow walkway, and exited under wooden scaffolding being built to cover the entire length of the Palace. Where the scaffold was already complete, architects stood on the wooden framework, scanning rolls of plans and arguing.

Dressed in their plush jackets and high heels, they looked out of place among the wild hair and smocks of the carpenters. Newton registered their presence but paid them little heed.

'I am told that the gardens have been neglected,' said the younger man.

'I don't think we need flowers to distract men like us, not when we have such conversation.'

Fatio blushed again. 'I am at your command, Mr Newton.'

'Lead on, Monsieur Fatio.'

Hooke watched Newton and Fatio from the scaffolding. He was still breathless from the climb and leaning against one of the uprights. 'Look at them, prancing and preening.'

'Leave them be. We have a job to do,' said Wren.

'Who's the pretty boy?'

Wren glanced up from the plans.

'That is Nicolas Fatio de Duillier, hero of the Dutch people.'

'Hero?'

'They say he foiled a French plot to kidnap William.'

Hooke stared at the receding men. 'He barely looks able to arrange flowers.'

Wren chuckled despite himself. 'Come along now, Robert. Let's get back to work. Have you recovered enough to finish the climb?'

Hooke's lungs were still burning, but he forced some indignation into his voice. 'Of course.'

Wren rolled up the plans and began climbing upwards again.

Hooke waited for him to dismount on to the higher platform and began his own ascent. Midway, his legs turned to lead and he slumped against the red wall for support. He snatched a glance towards the gardens. Newton walked easily, upright as a church tower, hands clamped behind his back. Fatio walked self-consciously, as if each movement were being judged. 'He walks like a woman,' Hooke called upwards before feeling for the next rung. He missed his step and the world dissolved around him.

When he came to he was lying on the grass, staring at the painfully bright sky. He jerked up into a sitting position. His head felt as if it had been split open.

Wren was crouching next to him.

'Easy, Robert.'

'Did I fall all the way?' Hooke looked anxiously at the tower of scaffolding.

'Not quite. Mr Trent carried you down.'

Cradling his head, Hooke sank back on to the grass and closed his eyes.

'I fear that this work is not for you any more, Robert.'

'I missed my step, that's all.'

'I know about your black-outs. They're getting worse, aren't they?'

Hooke had tried to keep them quiet, but perhaps Wren had heard about last week's incident. He had been caught unawares in Garraway's and sent an entire trestle table of coffee cups crashing to the floor.

'You're not safe to climb the scaffolding,' said Wren.

'But I'm younger than you by three years.'

'By God's grace, I'm still fit and able. I cannot risk you injuring yourself or worse.'

'I'd rather die working than fester in my bed.'

'You cannot do this any more, Robert. I simply won't take the responsibility. I'd never forgive myself.' Wren's words were softly spoken but his intent was clear. This was not a suggestion.

Gentle notes cascaded from the harpsichord in the gallery below the eaves. The servants lit the enormous round candelabra as evening fell and hoisted the giant wheels of light high, securing their ropes on hooks at the side of the dining-hall.

Newton was sitting next to Fatio and opposite Huygens. The Dutchman plucked food from his plate, talking between well-judged mouthfuls, but Fatio hardly touched a thing. Perched on his chair, he was leaning a little towards Newton.

The softness of the candlelight flattered the room's occupants, washing their faces with a gentle yellow glow. Newton stole frequent looks at Fatio's profile. Often he found the young Swiss man looking straight back at him.

'A toast to the new Master of King's College,' said Huygens, raising his glass to Newton. 'A letter will be drafted for despatch to Cambridge tomorrow.'

They drank the inky red wine.

Newton noticed the new King glancing towards him. He inclined his head to be respectful and thought he saw William return the gesture.

When Newton returned his attention to his immediate companions, Huygens was regarding him with a quizzical expression. 'I wonder if my admiration for you earns me the right to ask you one question, Mr Newton?'

'Of course,' said Newton, immediately on his guard.

'Your work admirably describes the effects of gravity and allows us to calculate the influence that one massive body will have on another, but it does not say how the gravity is communicated between those objects. What are your thoughts on this matter?'

Newton suppressed a stab of irritation. Huygens had found his Achilles' heel. He steepled his fingers and spoke with care. 'I have not yet been able to discover a single motivating reason behind the phenomenon of gravity. It is presently beyond my ability to understand.'

'But you must have some ideas. Come, don't be coy with us.'

'We're your devoted followers,' said Fatio, earning himself a sideways glance from Huygens.

'Gentlemen, I offer no hypothesis that cannot be proven. Yet . . .' The two foreigners leaned closer. 'Speaking confidentially, among friends, it is inconceivable to me that inanimate brute matter should operate upon other matter without mutual contact.'

'I thought so.' Huygens looked triumphant. 'You know, sir, that I have succeeded in compiling a mathematical version of Descartes's vortices.'

Newton swilled some wine. 'They cannot work.'

Huygens looked hurt. 'How can you dismiss them so casually?'

'You said that you had read and understood my *Principia*. Then you know that in book two I destroy the notion of vortices. Think of this. Descartes believes that the heavens are filled with an ether and that swirling eddies circle the Sun, carrying the planets. But if a vortex is carrying Jupiter on its orbit, then the planet's moons must also be caught. As they orbit Jupiter there must be times when they are travelling in the opposite direction to the vortex, yet they do not slow down as they should if they are travelling against this headlong wind.'

The Dutchman frowned. 'But without a mechanical explanation, your theory is incomplete.'

'My theory allows you to calculate the effects of gravity – what more do you need?'

'But it smacks of the occult. Without a mechanical explanation the forces must magically jump across empty space.'

Newton added what he hoped was a conciliatory tone to his voice. 'There are worse things to live with than a mysterious force of gravity.'

Fatio almost jumped from his chair. 'I can reconcile the two systems! Universal gravity with a mechanical cause.'

'How?' Newton indulged him with a smile.

'I would rather not say just yet, but I will be working on my own treatise.'

'Then you, sir, are a marvel to be toasted.' Newton wanted to reach over and touch the young man's hair. He distracted himself from such thoughts by raising his glass.

Huygens joined the toast. 'I told you that he has the learning. If anyone can do this, he can.'

Newton found that he did not mind the implied slight on his own abilities. Fatio's cheeks were flushed and his eyes sparkled. Newton said, 'If you can do this, I will have every Fellow at the Royal Society put his name to it. I've been thinking about your rebuff from Leibniz. You don't need him now, I will be your mentor.'

'Truly?' Fatio gasped.

'It will be my great honour.'

Fatio blushed, and Newton's heart soared.

30
Adriatic Sea

Heart pounding, ear pressed to the door, Gottfried Leibniz did not believe what he was hearing. The insistent voices were barely audible above the pitching ship's creaking timbers and the crack of thunder from the sky. When the voices grew fainter, Leibniz placed his eye against a knot-hole. He could see a pair of sailors lurching around, one of them pointing wildly in his direction as they climbed to the deck.

The vessel heaved again, reverberating under the impact of another crashing wave, and Leibniz fought the urge to vomit. Water dripped from the roof, cooling his face, and he forced himself to calm down.

His Italian was poor, but he had heard enough to know what the sailors were discussing. He knew the word *paria* when he heard it, and he knew what was done with such undesirables. On a night like this he would stand no chance in the rolling waves.

They had not been expecting the storm. It hit just as night had begun to lull their senses and the grog had minded them to sing their shanties. Gathered below decks, the first of the hammocks was being stretched across the beams when one white-haired mariner cocked his head. A moment later the first wave sent them sprawling, and while the crew scrambled for the deck, Leibniz returned to his cabin.

He turned from the knothole and tried to collect up the papers and cutlery that had been thrown about, but when he too was pitched into the bulkhead, he gave up and tried to lodge himself in a corner. The lanterns swung, toppling and extinguishing the candles inside. The only source of illumination was an occasional burst of lightning.

In his foetal position, growing ever more agitated, he tried to piece together the overheard conversation.

Lutheran . . . Omen . . . Sinner . . .

He was sure that the angry sailor was the same one who had been eyeing him with suspicion ever since he had boarded at Venice.

'*Tedesco*?' the sailor had said that first night. '*Tedesco*?'

'I'm German, yes. On my way to Rome.'

The man had pulled at the lower lid of his eye and spat at the philosopher's feet. Now he wanted to throw him overboard.

As another wave struck the ship, panic surged through Leibniz. He tore open the cabin door. Where could he hide? Down in the hold perhaps, with the cargo? Surely they would know the nooks and crannies better than him. So where? Where could he hide?

Over the noise of the storm he thought he heard shouting from the deck; certainly there were feet stamping on the single wooden layer that separated them. They were coming back. He was sure of it. Throw the Lutheran sinner overboard and God's wrath would abate.

Then he saw it: a string of beads in the darkness. They were rolling around the floor, probably having fallen from a sailor's pocket as he scrambled to save the ship. Leibniz snatched the beads and headed back inside.

Kneeling in the corner, it took him no effort to pray. He mumbled a litany and fed the beads through his fingers one at a time.

He jumped when the sailors burst through the door but forced himself not to look up. He was handling the beads for all to see and muttering his prayer. Lightning flashed and the mob froze in the white light. They stared at him in silence.

'*In nomine Patris, et Filii, et Spiritus Sancti,*' he said loudly.

'Amen,' one of them responded, crossing himself.

Another one took a determined step forwards but a burly fellow with drenched hair and beard grabbed his shoulder. '*Signor*,' he said in apology and backed out of the room, pulling the troublemaker with him. The others followed.

Leibniz kept up his prayer for a long while afterwards. He held the rosary beads all night but did not close his eyes to sleep. When the battered ship eased into the next port he slipped away and continued to Rome on land.

The piazza was full of people. A sudden shower propelled Leibniz into a shop doorway. As he sheltered he watched the citizens go by: unconcerned by the change in the weather, they dodged around each other, eager to be about their various journeys despite the rain.

Hawkers still called out their wares, eager to make some fast money from ripe fruit, and it seemed to Leibniz that people were always on the move in Rome. It was so different to life back in Hanover, where citizens braved the cramped and muddy streets only as a last resort.

When the shower died away he set off again, bound for the enormous Jesuit Roman College. Situated on the opposite side of the square, stone walls glistening, the towering structure provoked envy in him. He could not help but admire the Catholics' skill of organisation, of bringing together scholars, philosophers and theologians under one roof. He imagined the debates and the discussions stretching into the night as the college reached its consensus. Much better, he thought, than the piecemeal approach in northern Europe, with individuals vying for status and philosophical power. In London, they at least had the Royal Society; frankly, anything would be better than the philosophical isolation of Hanover.

Making his way through the crowd, Leibniz felt out of place. His woollen jacket was far too warm for Rome, and despite the shower his raven-coloured wig was a cauldron on his head, drawing sniggers everywhere he went. He was taller than most Italians, and the towering hairpiece was not helping matters. He had tried going out without the cursed thing but his naked scalp embarrassed him. A Turkish headscarf had attracted the most blatant stares of all. In the end, he decided the occasional laugh at his periwig was the most bearable option.

As he entered the Roman College, his envy of the Jesuits increased. He stood overwhelmed by the palatial lobby, his gaze drawn by the towering Egyptian obelisk in its centre. It must have been twenty feet high, yet it sat as comfortably as a potted plant in an ordinary home. It was covered in ancient, undecipherable markings.

'They say Galileo himself stood and looked at it on his first visit to this college,' somebody said in passable German.

Leibniz turned. The voice belonged to a youthful man with wide eyes and a broad forehead.

'Don't look so shocked. We're allowed to mention Galileo, you know, just not his science.' The Jesuit grinned.

'Father Grimaldi?'

'Gottfried Leibniz, I'm so pleased to meet you.'

'Thank you for inviting me here today, Father. There are many who would not have wished to be seen with a Protestant historian from the court of Hanover.'

'You're a historian now? I know of your work as a philosopher, a mathematician, even a theologian, but not as a historian,' Grimaldi teased.

Leibniz winced. 'I do whatever the Duke requires of me, and at present he would like me to trace his ancestry as far back as Charlemagne. That means travelling to find the covenants and birthrights. But at least it means I can indulge my passion for philosophy with people such as yourself.'

Grimaldi bowed his head. 'You're most kind. Now, please, let me take you for some refreshment. You look rather hot.'

Leibniz mopped his brow. 'Delighted.'

Grimaldi's office looked out over a landscape of orange and grey slate. There was no trace of the rain now, the clouds having evaporated along with the moisture from the thousands of roof tiles. It was impossible to see the streets, as the buildings appeared to lock into one another, and Leibniz had the sensation that he could step out and walk across the roofscape.

Grimaldi handed Leibniz a goblet that smelled of peaches. Entranced by the exotic flavour, Leibniz downed the cordial in one gulp and smacked his lips with appreciation.

His host's eyes creased with pleasure. 'They grow them in orchards just south of here, the same family for seven generations now.'

'Delicious, but small talk is difficult in a foreign language, and I don't think that sampling peaches is why you asked me here today.'

'I like your directness. It's a welcome trait around here.' Grimaldi led Leibniz away from the view to a small, battered desk. The Jesuit squeezed behind it and sat down. 'Now, let me be direct with you. Rome is in philosophical crisis. It's nervous about the new philosophies coming from the North. Should it be?' Grimaldi pulled open a drawer and placed a copy of the *Principia* on the table. A number of paper strips had been interlaced as bookmarks, with notes scribbled on them.

'To me, Newton is one of the few people alive who is pushing the boundaries of science. His grasp of mathematics is superb . . .'

Grimaldi leaned in. 'I sense a *but* coming.'

Leibniz eyed the Jesuit. 'There is something dark, maybe even dangerous about his ideas. I believe they may even be irreligious.'

'Explain,' said Grimaldi, intrigued.

'Nowhere in the work does Newton tell us what gravity is. He states that it is merely action at a distance – some sort of force that appears from nowhere and travels through space. It is a concept taken from the alchemists.'

'Go on.'

Leibniz continued. 'Some years ago I tricked my way into an alchemical society. I was curious about alchemy – the secret nature of it fascinated me – and I soon heard rumours of an adepts' society in Hanover. But they'd only admit other adepts, not young men eager to learn. So I bought any old alchemical manuscripts I could find, determined to decipher them. Of course, it was useless. I couldn't understand a word, but I didn't let it stop me. I wrote a paper about alchemy, using the most complicated of the words and phrases I had read. Have you read any alchemical writings?'

Grimaldi shuddered. 'I have always avoided them.'

'They're full of allegory and metaphor. They read like veiled myths, which the alchemists believe contain clues to the lost knowledge of ancient civilisations. So I used the most extravagant of their language and wrote something that sounded meaningful but was, in truth, gobbledegook. Within a week, I was approached and invited to a meeting. I learned that they believe all things, from the stars to the animals, are governed by a universal spirit. This spirit gives rise to what they call *active principles*, which cause change and transmutations on Earth. They believe that if they learn the secrets of these active principles, they will have the power of God.'

Grimaldi's face turned ashen. 'But what has this to do with Newton's *Principia*?'

'As I said, nowhere in the book does Newton provide an explanation of what gravity is. In fact, he goes out of his way to try to show that there can be no mechanical explanation. He attacks vortices, even the concept of the ether. What does he advocate instead? Active principles. May I?' Leibniz indicated Grimaldi's copy.

Grimaldi pushed the book towards him.

Leibniz found the page and read out the Latin. '*A certain most subtle Spirit pervades and lies hidden in all large bodies. By the force and action of this Spirit, the particles of such bodies mutually attract one another. But these are things that cannot be explained in a few words, nor are we furnished with that sufficiency of experiments which is required to an accurate determination and demonstration of the laws by which the Spirit operates.*'

Leibniz replaced the book on the desk. 'Thinking such as this comes directly from alchemy. Newton is a secret adept, and this is an attempt to foist occult forces upon natural philosophy – and justify them with mathematics.'

'I had no idea.' Grimaldi leaned backwards in his chair. 'I thought he was glorifying the Holy Spirit.'

'To an alchemist, even their state of mind can override God's will and change the outcome of their experiment. They search for power over nature, not simply an understanding of it. Newton may be trying to pervert us all.'

Grimaldi was looking at the book as if it were cursed. 'I have never before felt the desire to burn a book.'

'Don't burn it just yet. As I did with those alchemists back in Hanover, I intend to understand what Newton's motivations are. His mathematics is sound; his description of gravity's behaviour is a masterpiece of reason. It's the justification he places on top of it that's questionable. There's plenty to learn from Newton, but, as a sinner must be cleansed, so we must choose what is pure about the *Principia* and exorcise the rest.'

'Are you up to the task?'

Leibniz met his gaze. 'I am.'

31
London

Halley had already received one drenching that morning and was in no mood for another. He glanced skywards, but the grey weight of clouds was making no promises. Returning his eyes to the muddy ground, he dimly perceived a hatted figure leaning nonchalantly on the stone pillar of the iron watergate at the end of the street. The faint lap of the river could be heard beyond.

Halley rapped on a narrow black door.

He was greeted by the housekeeper as he scraped the mud from his shoes on the boot-scraper set into the ground. Once he was inside, Mary Skinner, Pepys's companion for the last two decades, hurried out of the drawing-room.

'Edmond, you've had a wasted journey. Sam's not here, he's been arrested,' she said, more annoyed than upset.

'Arrested?'

'They say he's plotting for James's return.'

'Is he guilty?'

A thunderous look crossed her face, 'Even you doubt him?'

'I didn't say that.' Halley silently cursed his ineptitude. 'It's just that he was on the very best of terms with the previous King . . .'

Mary pushed her fists into her fleshy hips. 'He served all his Kings to the letter. He was no traitor.'

'I'm sorry, Mary. That was clumsy of me. What evidence are they using against him?'

'Nothing.' She shook her head. 'It's all contrivance.'

'Then he'll be home soon, I'm sure of it. He's been a good friend to me.'

'If Sam ever wished to scare me, he'd tell me about Tangiers, and how there were parts of Africa where the sands look safe to walk on but would suck you down to your death. I fear England's become the same.'

Halley risked a brief touch of her arm. She did not respond. 'These things will resolve themselves. Is there anything I can do?'

She shook her head and bade him goodbye.

Halley set off at a determined pace, one eye on the scudding clouds above him. At the end of the street he was passing an awkwardly parked carriage when the door swung open. A brass-topped walking cane emerged from the door at chest height and brought him to an abrupt stop.

'What's the hurry, Edmond, old friend?'

'Mr Winslow.'

'It's been too long. Get in.'

The large man with the black hat came up behind Halley, leaving him no choice but to comply. The oaf heaved himself in too, making the carriage tilt.

'I've nothing to tell you,' said Halley, recalling the questioning after his trip around Europe.

'Oh, I think you have.'

His abductor held out a black hood. 'Put this on, please.'

Halley stared defiantly. 'Is that really necessary?'

The giant next to him stirred. Halley reached out and placed the hood over his head. He could feel the drawstrings being pulled tightly around his neck. The carriage lurched into motion for a while, then shuddered to a halt. He was manhandled out of the vehicle and told to walk.

Indoors, the hood removed, Halley found himself in a dark room. A thick tapestry hung at the window, a few thin needles of light penetrating the hunting scene. An iron candelabrum stood in one corner, supporting three large candles that flickered their meagre light around the room, and a perforated wooden screen stood in another. Halley thought he could see a silhouette behind the screen.

Winslow sat opposite, a table between them. The deep shadows produced by the candlelight emphasised his emaciated appearance. He unrolled a printed sheet of paper on the table. 'Do you recognise this?'

'Of course. It's my survey of the sandbanks in the Thames estuary.'

'And you're proud of the achievement.'

'I am. I hope I have saved lives by making the approach safer.'

Winslow pulled a sour face, unrolling another sheet. 'And this?'

It was Halley's southern star chart. 'How did you get that?'

Winslow leaned closer. 'Take my advice. It's best not to ask questions in situations like this. You dedicated the chart to Charles II, a bold move for one as young as you were then, don't you think?'

'He was my monarch at the time,' said Halley, beginning to feel uncomfortable.

'As was James when you made the estuary survey. How well do you know Samuel Pepys?'

'I know you've arrested him. I went to visit him today, as you witnessed, so I presume I'm under suspicion too. Pepys and I have the Royal Society and the Admiralty in common. That is all. I'm English to the core and I will serve the English monarch.'

'Even if that monarch was intent on leading the country back into the arms of Rome?'

A wave of irritation swept through Halley. It was as sudden and complete as the drenching he had received from the rain earlier. 'I am a loyal servant and wish to serve in peace. And I'm in no mood to be lectured by you,

Winslow. You served Charles and James much more closely than I ever did.'

Winslow momentarily bared his crooked teeth. He glared at Halley, but not before his eyes had flicked to the wooden screen.

The shadow behind it moved and the realisation struck Halley with full force. He whirled from the chair and dropped to one knee. 'Your Majesty, I wish only to serve you as best I can.'

'How dare you–' began Winslow.

The King emerged from behind the screen. Halley, head bowed, saw a pair of garishly buckled shoes encasing puffy ankles.

'I have heard enough,' the King said in his Dutch accent.

Halley kept his head bowed and said nothing.

'We must all adapt to the world in which we find ourselves,' said the King. 'Stand up, Mr Halley. I have heard of your father, his association with the Earl of Essex and with my father-in-law . . .' He hesitated as if he didn't quite believe his own connection to James.

Halley rose. The King had sharp features and chiselled lips. His cheeks were rouged but there was fairness in his eyes, perhaps even kindness. He was swathed in a cloak of lush red velvet.

'I do not believe you would want James back on the throne. Go back to your telescopes and serve in peace. We may yet call on you in the future to serve your country again, but there has been enough tyranny in England.'

With a shiver of excitement, Newton opened the door to Fatio. The young man stood in the last of the Cambridge daylight. His head was held high, yet he grew bashful when accepting the invitation to step inside.

Newton stepped aside, but still their bodies passed within inches.

The waft of fresh air reminded Newton just how strongly the room smelled of hot metal.

'What is this place?' Fatio stared into the gloom at the angry light from the furnace.

'Have you never seen an elaboratorium before?'

'Not like this.' Raising a hand to his nose, Fatio scanned the rows of flasks, alembics, crucibles and chemical jars before approaching the furnace and its bubbling flask. He turned to Newton, wide-eyed. 'You're an adept?'

Newton spoke softly. 'Some call me that.'

Fatio pointed to the flask. 'What are you making?'

'Something especially to show you. Watch carefully.'

Newton lifted a pair of tongs and grappled the flask on to a metal frame. 'Observe what happens when it cools.'

In the flask, tawny particles were condensing out of the fluid. They clung to the sides and the flat bottom of the glass, sprouting tendrils that grew quickly through the liquid. Newton had seen it happen countless times, yet he still found it fascinating.

'They look like plants growing at supernatural speed,' whispered Fatio, leaning closer to the transformation.

'These are the active principles of nature at work.'

'Is it alive?'

'Not as we are, but it must be tapping some measure of the same animating force. Look at the organic structures. Whatever this force is, it must flow through all matter, ready to be used.'

The transformation was slowing, having filled the flask with a tangle of chemical branches.

'What's it called?' asked Fatio.

'The adepts refer to it as the regulus of Mars. It's based on antimony. It's an important step towards the Great Stone.'

'The Great Stone?'

'The Philosopher's Stone. Those who know little of its power think it is simply to be used to create gold from lead, but this' – he indicated the crystalline briar – 'is the real power of it. Nature delights in transformation. All around us we see things changing into other things. The Philosopher's Stone would allow us to control those transmutations. Lead into gold is nothing compared to its true power.'

There was a trace of fear in Fatio's eyes, and Newton realised just how immature the man was. He smiled to soften the moment.

'Gravity is just the beginning,' he said quietly. 'It gives us the movement of the gross collections of matter, but there are other forces at work too, threading all matter. That is my ultimate goal: the philosophical understanding of everything. I'm close to it – I can sense it. I've felt for a long time that some kind of judgement is hanging over mankind, but once I have the power to control transformations I can end all the fear and uncertainty.'

Fatio looked as if he might cry.

'But I see I have scared you. Think no more of it. These are indeed fanciful thoughts, and sometimes I get carried away. Tell me of your theory about the cause of gravity.'

Fatio bobbed his head. 'I find myself reticent.'

'Come, now,' Newton purred, 'what secrets can true friends have from each other? I have confided in you.'

Fatio composed himself. 'I believe that the ether is in constant motion, tiny particles all rushing hither and thither. There are neither vortices nor streams; the particles travel in arbitrary directions. Here, let your fingers be the particles.' He waggled his fingers at Newton, daring the professor to do the same. Newton lifted his hands and joined in. Something that might have been amusement or embarrassment lodged in his throat.

'Now, if a planet were to be suspended in space, these particles would collide equally from every side, the forces would cancel and the planet would remain still.' Fatio raised his fist and Newton played his fingers over it, feeling how warm the skin was.

'Put another world nearby and it shields the first from a particular direction.' Fatio raised his other fist close to his first, forcing Newton's hands apart so that his fingers spread themselves over Fatio's fists.

'Now the forces of collision are no longer equal and the planets will be pushed together.' Fatio began to draw his hands together as if Newton's dancing fingers were forcing them to move. Newton followed the motion, ending up with his hands encircling Fatio's clenched fists.

Newton's mouth went dry. With an effort of will he released Fatio's hands. 'Ingenious. We must get you on to the agenda at the Royal Society.'

'Truly?'

'Why not?' Newton exulted in the look of triumph his praise evoked.

Some weeks later, Newton and Fatio dodged the carts trundling along Bishopsgate and headed for Gresham. Fatio eyed the crumbling red bricks; one of the many chimneys had collapsed to make a small pile of rubble in the street and leave a hole in the moss-ridden roof.

'*This* is the Royal Society?'

It was indeed an embarrassment for the Society to meet amid such ruin, but Newton was not prepared to admit it. 'The Fellows are the Royal Society, not the building, but don't be intimidated by them. They will look at you with blank faces and close their eyes, and Hooke will probably pick his nose, but don't let them worry you. You're better than any of them, even Halley – though don't let him know I said that. Not yet, anyway.'

'If you say so.' Fatio adjusted his jacket collar and stepped into the peeling entrance.

In the quadrangle, laundry hung from windows. There were shadowy figures in the cloisters.

'There's Robert Hooke,' said Newton.

The hunchback was talking to two young men. Both were tall and slim, and dressed in such similar cuts and colours that Newton thought they could be twins.

'But we must talk to you about the tenancy agreement,' one was saying.

'The tenancy agreement,' echoed the other, a half-beat behind.

'Not this again. Tell the governors that I will not move out.'

'Please see reason, Mr Hooke.'

The men were so tall that Hooke's eye-line was level with their lacy neckcloths. He stepped back. They followed. 'Stand off, sirs, so that I may see your faces.' There could be no doubting his peevish mood. 'There's nothing wrong with this building. It's stood for more than a century and it will stand for another. It will not be torn down now on a whim.'

'But the building is so outmoded, and it is largely unoccupied.'

'Yes, unoccupied.'

Hooke remonstrated. 'The professors are still here. If you and the trustees cannot let the rest of the rooms, it's hardly a failing on my part. This is my home. I have given everything to this college, made it not just my home but also that of the Royal Society. The *Royal* Society. Do you want the French to overtake us in matters of natural philosophy?'

The two bailiffs exchanged mystified glances.

'Of course you don't. Now, gentlemen, if you will excuse me.' Hooke pushed in between them.

'Mr Hooke,' Newton said. He planted his words with enough fake alacrity to let Hooke know he had witnessed the attempted eviction.

Unexpectedly, Hooke's grey cheeks rose in a smile. 'Solved the movement of the Moon yet?'

'Oh, not quite.'

'Flamsteed tells me you've been begging for observations again. It must be a real cross to bear when your theory cannot solve a simple thing like the movement of the Moon.'

Newton strode past the professor, Fatio trailing behind. 'I take it your chambers are open.'

There was an embarrassed silence in Hooke's draughty sitting-room as Fatio drew to his conclusion. The men pulled their chins, studied their feet, did anything to avoid making eye contact. Newton's blood boiled. He stood up, deliberately rattling his chair.

'Congratulations, Mr Fatio. One of the most promising lines of reasoning I have ever heard.' He disregarded Halley's look of surprise and said, 'I'm sure the other Fellows wish to join me in congratulating you.'

There was a faltering round of applause, which soon died away.

'Fellows!' called out Halley, bouncing to his feet. 'Before you depart, I have an announcement. We must celebrate the engagement of our very own John Flamsteed to Margaret Cooke, spinster of Barstow – and eight years his junior.'

The Astronomer Royal crossed his arms, displaying knuckles swollen with arthritis. 'You are a wretched creature, Mr Halley. You belittle the sanctity of my marriage with your inappropriate comments.' His crisp enunciation conveyed no humour.

'But I meant no harm. I just didn't know you still had it in you . . .' The joke died on Halley's lips.

Flamsteed rose. 'You drink too much, you swear too much, and you have too much of an eye for the ladies.'

'Say what you like about my drinking and swearing, but I am happily married.'

'I see now that my early faith in you was woefully misplaced.' Flamsteed headed stiffly for the door.

'Mr Flamsteed, before you go' called Newton. 'You have been kind enough to send me observations in the past.'

'I'm always willing to help a fellow philosopher.'

Newton stifled a snort of derision. *Flamsteed was trying to pass himself off as a peer. Still, needs must* . . . 'I find I must ask for more to propel my lunar theory to completion. It strikes me that such a theory will be the perfect proof of your observing skill. It will show everyone that you are the most accurate observer of the heavens who has ever walked God's Earth.'

'Thank you, Mr Newton. I will send you observations and my own lunar computations, which I believe you will find most useful.'

'Oh, I'm not interested in your computations,' said Newton. 'All I require are raw, accurate observations. I found a number of mistakes in your last data.'

Flamsteed looked deflated. 'Mistakes?'

'Yes, mistakes. Figures so wildly divergent from the others that they can only be mistakes. Please check them next time. I must have accurate figures. And keep your computations, I will do the rest. I will even send back my computations so that you may use them in the completion of your star catalogue.'

Flamsteed gaped.

'Do you have nothing to say? Let me know, plainly and honestly, whether this arrangement pleases you or not.'

'I think I do not like this arrangement.'

'Why hoard your observations when they can be useful to me now?' Newton growled. 'Let me prove your reputation, because so far, Mr Flamsteed, you have published so little that you have given no one any reason to believe you are actually achieving anything at Greenwich.'

Flamsteed's face contorted. 'Very well, Mr Newton, I will tell you plainly: I choose to publish my observations in my own way, in my own time, when I am satisfied with them. And you can see them at the same time as everyone else. Once more, I bid you a good evening, gentlemen.'

'That's both of us in his bad books now,' said Halley with a grin.

Newton ran his fingers along the cold lead of the window frame in Fatio's sitting-room. Even with his jacket wrapped around his shoulders, he felt the chill of the night. Outside, London slept beneath the Moon.

He could sense the solution to its orbit, as if the equations were in the corner of his vision, perceptible but lacking detail. He knew he could not finish the computation without accurate data. It was a scandal that Flamsteed had not yet published a catalogue after working for more than twenty years.

Newton pressed his palm against the glass. His head was filling with chatter again: perhaps a new mixture for the furnace, or a fresh interpretation of a biblical passage. He struggled not to waste these fleeting gifts of insight, but there was only so much that he could do.

'You're still up.'

He looked round towards the voice. Fatio was wearing shirt and hose, his hair disarranged.

'Sleep evades me,' said Newton.

'Is your bedroom not to your liking?'

'It's fine, thank you.'

'Cognac, that's what you need.'

'I wish you wouldn't drink so much,' Newton said tentatively.

'I wish you wouldn't worry. A little drink is good for the nerves.'

Newton found the naivety painful but could not bring himself to correct the boy. 'I feel trapped by my work, maybe even a little apprehensive.'

'You? Why?'

Immediately Newton regretted the confession. How could he make Fatio understand? The blackness of those desperate times writing the *Principia*; the torture of trying to make sense of all the revelations. How could he go through that again? Yet how could he stop now?

Newton turned back to the window. There was another fear gnawing away at him. 'You're a young man, Nicolas. I fear marriage may take you.'

'I will never marry for anything other than love, Isaac, and I think there is a reason at work in my very soul that means I will never marry.'

Newton searched Fatio's face. 'What is that reason?'

'I think you know.'

At that moment Newton knew of nothing beyond an impossible longing for the young man in front of him.

Fatio put down his glass and embraced his friend. Newton's body reacted so powerfully that he began to shake. He gripped Fatio, surprised at the insubstantial nature of him. He relished the intoxicating sensations for a moment, then fell back as he realised how obvious his arousal must be to Fatio.

'There is nothing to feel ashamed about, Isaac. Nothing improper has happened between us . . .'

But Newton could hear the unspoken 'yet' in Fatio's voice. He composed himself. 'John Locke has returned to England. I will be leaving to visit him in the morning; Mr Huygens has put us in touch, as he promised he would.'

'You didn't say.' Fatio looked hurt.

'I didn't want to upset you before your presentation today.'

'May I come with you?'

Newton turned his back. 'I don't think that would be a good idea.'

'Why not?'

'Mr Locke and I have things to discuss that you will find tedious.'

'I see.'

There was a rustle of movement. When Newton turned around, Fatio had disappeared.

32
High Laver, Essex

John Locke was a stick of a man. Swaddled in donkey-brown robes, he wore a high-necked white chemise of the type worn by learned men to have their portraits painted. 'It's pretentious of me, I know,' he confessed, leading Newton through to the back of the house, 'but I've grown rather fond of their comfort.'

Topping Locke's small body was a heart-shaped face that seemed to exaggerate the size of his brain. His hair was a silken bonnet of white and his deep-set eyes projected a doleful acceptance of all he regarded. He moved carefully, but not inflexibly like so many men of his age.

'It's good of you to come and visit me out here.' His voice was soft but strong. He showed Newton into an austere room that overlooked a terrace and a garden ringed by trees. Beyond the lawn his visitor could see an extensive vegetable patch.

'You're no further than Cambridge,' said Newton, 'and I'm getting rather tired of that journey.'

'I heard of their refusal to accept your appointment as Master of King's College, despite the King's express wish.'

Newton waved the matter aside. 'There are other positions far from Cambridge, but I'm not here to talk about that. Of all the letters I have received about the *Principia*, yours have intrigued me the most.'

'Then I am honoured. I thought my inability to follow your geometry would send me to the bottom of the pile.'

'There are more things than geometry that interest me.'

'Indeed.' An ambiguous expression crossed Locke's face.

There were few pieces of furniture in the room. It could have passed for a monk's cell if not for the glass-paned doors that led out on to the terrace. A maidservant entered and handed them glasses of cordial. Locke waited several heartbeats after her departure before saying, 'I think that one of the things we can agree on is that religion can take many forms, even within the same Church.'

Newton sipped the sweet drink, forcing Locke to continue.

'I have advocated a form of religion that stresses reason above emotion, because I think in the fervour to promote one's belief heady emotions can lead to great injustices.'

Newton gave a slight nod.

'Think of the tragedy of Galileo,' continued Locke. 'Such destructive emotions directed against him when all he spoke of were his observations. Absolutism in the scriptures can be as dangerous as absolute monarchy, do you not agree?'

'Religion should be one's armour, not one's sword,' Newton said.

'Quite so, yet I fear others may suffer his fate unless we can change society for the better. We need reason to guide us, not passion, and to do that observations are paramount. We must see the world as it really is, not as we imagine it to be. Even the lowliest beggar can imagine a world filled with food and comfort, but that doesn't make it happen because he has no power to change the world. A King can make things happen because he has people and money at his disposal. So these two individuals, of such different stations in life, are unlikely to agree on whether life is equitable. Yet turn to nature, and both beggar and king can agree on hearing the call of a blackbird on a summer's evening, or seeing the brief fire of a meteor. In our observations of nature, we find something that every rational man can agree upon.' Locke's eyes fixed on Newton.

'I agree,' said the professor.

'Your method of observation, hypothesis and experiment would seem to me to be the way to guide our society to a more rational state of being. We've taken good steps recently with the abolition of the divine right of monarchy. I venture that constitutional monarchy under the arbitration of Parliament is the most important step towards the perfection of England since Henry Tudor's split with Rome. Never again will a king or queen be able to hide behind God's grace to justify intolerable actions. It's a great step towards rationality. We need to take others, of course.' Locke's expression intensified. 'I don't mean to patronise you, Mr Newton, but your new method for investigating nature. .. I wonder whether you comprehend how influential you could become? A new philosophy, to bring us to a true understanding of God.'

Newton was unbalanced by Locke's gaze. 'You keep an austere home,' he said, trying to buy time.

'It helps me think clearly. It's too easy to end up placing too much value on things, do you not think? Pictures? Sculptures? We fail every time we value *things* too highly.'

Newton picked his words carefully, grasping Locke's intent. 'Clutter. I sometimes think that the Second Commandment has been rather long forgotten.'

'Idolatry,' agreed Locke. 'A problem, I believe, for centuries.'

'Since the fourth century?'

'The defeat of Arius.' Locke stepped closer. Newton followed. The two men clasped each other's arms.

'I thought I was the only one,' said Newton.

Locke tightened his grip. 'Rest assured, you are not alone.'

The forest air was perfectly still and cool enough to be refreshing. It pushed through Newton's lungs, loosening his arms and shoulders. Soft moss cushioned his feet and the pungency of wild thyme filled his nostrils.

'When did you know?' asked Locke as they walked.

'At the height of my battle with Hooke over the origin of colours. His ignorant bluster disturbed me so much that one evening I turned to the Holy Book for solace. As I was reading Proverbs I found a paradox. Christians believe that "The Lord created me at the beginning of his work" refers to Jesus. If that is the case, then Christ cannot be as fully divine as the Father, and so the doctrine of the Trinity is false. I saw it as clearly as if a blindness in me had been cured.' He could still recall the dagger of fright that had pierced him at that moment. 'I read on, looking for my error, but found only confirmation: "There is one God and one mediator between God and men: the *man* Jesus Christ"; "The head of every man is Christ, the head of every woman is the man, and the head of Christ is God"; "He shall be called the son of the most high",' quoted Newton. 'Within a matter of days, my faith in the Trinity had been completely undermined. So I turned to the theologians who justified it and found only perversions of the truth in their words.'

'We have taken similar paths,' said Locke, his face dappled in the light.

'Imagine how I have lived with the horror of this knowledge ever since – a Fellow of Trinity College! My dilemma reached its crisis when the Senate began reminding me that I was approaching my deadline for ordination. How could I become an Anglican minister when I no longer believed that the Father and the Son were of the same divine substance?'

'And that's why you went to petition the King in London.'

'You're well informed.'

'I heard of your stand before Jeffreys. What I cannot fathom is how you persuaded the King to agree to your dispensation. You can hardly have revealed your views.'

'I let him know that I was bound to an older doctrine than that of the Anglican Church.'

'And he assumed you meant Catholicism.' Locke's eyes widened. 'You deceived the King of England?'

'I spoke nothing but the truth.'

'Extraordinary!' Locke shook his head in admiration. 'We have much to discuss, including how we are to progress our thinking. Will you stay the week?'

Newton hesitated. Fatio would be expecting him – or perhaps not, after their fractious parting. He felt guilty as he recalled their argument. He turned to Locke. 'I'd be delighted.'

In the days that followed Newton grew to love the simplicity of the house. Locke was right; it was perfect for contemplation. The house library was as well stocked with theology as Newton's own, and, in deference to the bare working room, Newton rationed himself to taking in one book at a time.

One afternoon, looking out at the wagtails chasing unseen insects across the lawn, he found his mind drifting back to the plague summer of 1666, when Cambridge had been abandoned and he had returned to Woolsthorpe and his mother. That had been the last time he had enjoyed this sense of unbridled scholarship – more than twenty years ago.

That evening, as he and Locke sat drinking coffee, he asked, 'How many like us are there?'

'Not enough, yet. I know of only a small number, but there is a way for us to test who else may be sympathetic.'

'How?'

'A pamphlet. Write me a simple argument that presents the evidence.'

'Publish? I should be damned as a heretic.'

'I can arrange for it to be anonymous.'

'But I risk catastrophe.'

'I'll send the manuscript to friends in the Netherlands, have it translated into French and published in France. The

French will no doubt flare up, thus guaranteeing its being brought to the attention of all countries. We sit back and gauge the reaction.'

Newton felt as if he had fallen into a trap. 'Is this the reason for inviting me here?'

'Look how far you've moved already: *Principia*, the constitutional monarchy. Surely the next step would be the return of the true Church?'

Courage and fear battled inside Newton. He had long tried to imagine this moment. He had envisaged it occurring later in his life, after he published his final treatise on natural philosophy, but perhaps now was as good a time as any. He was forty-nine; time was no longer his greatest ally, although – God be praised – he had yet to feel any pangs of mortality.

Yet the price if it were to go wrong! Even gravitation could be rejected if his reputation were ruined. For a moment he regretted making the *Principia* so difficult to understand. It made it all the easier for others to ignore.

'Will you do it?' pressed Locke, his doleful eyes unblinking.

Newton's heart pounded. 'To worship Christ as if he were the Lord is to me the foulest notion.' Closing his eyes he said, 'Yes, by my God-given strength, I will do it.'

On the fourth day, Locke and Newton returned from a walk at dusk and took their seats for supper. The pair sat at a table fit for twenty, in a narrow hall slung low with tarred beams. Locke was eating his usual morsels of vegetables while Newton's stomach roared for meat and hearty sustenance.

A servant appeared. 'A visitor insists he sees you, sir.'

Locke wiped his mouth with a cloth and made to get up.

'Actually, sir, I was addressing Mr Newton.'

Fatio stepped into the room. He was wearing a flamboyant new green jacket, piped in gold, and an expression of defiance.

A shadow of suspicion crossed Locke's face.

'Monsieur Fatio de Duillier is the most promising mathematician of his generation,' said Newton around his mouthful. 'We've been working together in London.'

'Then, Mr Fatio, you're welcome. Join us for supper,' said Locke coolly.

Working in the spartan back room, Newton ignored the first knock at the door. In his experience, that usually deterred unwelcome visitors. Not so in this case. The door hinges squeaked and a figure edged inside. Newton did not look round but hunched low over the work-table. It was hardly big enough for a single sheet of paper; the inkwell and finished sheets had to be laid on the floor, but it sufficed. This tactic of engrossment also often worked, compelling the intruder to retreat.

'I wondered if you would like to walk with me into the village?' said Fatio, though his tone implied it was not a request.

Newton finished writing his sentence. 'It is not yet time. We usually walk after lunch, to aid the digestion.'

'But I thought you might like to walk with me alone.'

'This is a most painstaking document.'

'You would go if Mr Locke asked you.'

Newton swung his head. 'Nicolas, do not mistake my being busy for not wanting to be with you. Sit with me while I work.'

'No, thank you. I have things of my own to do.'

At lunch, Fatio remained silent as Newton and Locke discussed progress in guarded terms. When the trio set out for their walk, Fatio sullenly tagged along behind, paying undue attention to the plants.

'Tell me, Mr Newton, does your study of nature bring you closer to God?' asked Locke.

Newton swept his gaze across the forest. 'Indeed it does. Nature does nothing in vain. Its transformations bring forth all the order and the beauty in the world. When I look at it, I'm convinced that there is an intelligence behind it.'

'Quite so, but there is one question I cannot resolve. Your laws of motion and gravity, they imply that the Universe is clockwork, running on mechanical principles. If that be the case, I am unable to see the role of God following the initial creation.'

'Then let me explain to you about the sensorium of God–'

Fatio suddenly uttered a succession of guttural, anguished sounds.

Newton turned to see him doubled up, clutching his stomach. 'Nicolas, are you ill?'

'I'm strangled in the belly.'

'Was it something you ate?' Locke inquired.

'Let me get you back to the house,' said Newton.

'No, you are deep in discourse. Do not let me stop you. What can I add to your elevated discussion?' Fatio turned and limped back along the path.

Newton did his best to return to the conversation, but a dozen steps later he halted. 'I'm sorry, John, I'm concerned about Nicolas. I must attend to him.'

'Of course.'

Newton expected to catch up with the pain-stricken Fatio, but the man was nowhere to be seen. Back at the house, Newton headed for the kitchen.

'Has Monsieur Fatio returned?'

'Monsieur de Duillier is in his room, sir,' a servant replied.

'I understand he's ill.'

'He seemed quite well when I passed him on the landing just now, sir.'

Newton bounded from the kitchen and burst into Fatio's room without knocking. The young man was sitting dejectedly by the window.

'What is going on?' asked Newton.

'I've been waiting for an opportunity to see you alone. I bring important news.' Fatio held out a worn leather-bound book.

'That's my Bible,' said Newton.

'There's a message in it,' Fatio weighted his words, 'from Reverend Flamsteed. He called the day after you left.'

'Flamsteed?' He closed the door and took the Bible. There was a strip of paper sticking out of the top. On it was written: *Read Jeremiah, chapter 10 to the 10th verse.*

Newton did not need to, especially not after the work he had been doing this week. Those were passages against false idols and liars.

'What does it mean?' asked Fatio.

'Did he say anything else?'

'No, what does he mean?'

'It's nothing you need to worry about,' Newton said distractedly.

'Isaac, please. Secret meetings, cryptic passages, veiled conversations with Locke; what's happening? Don't treat me like a child. I thought we meant something to each other.' His young face was hot with indignation.

The sight pierced Newton. He sat on Fatio's unmade bed. 'Come and sit down. I'll explain.'

Fatio sat next to Newton.

'In the fourth century there was a fierce debate in the Church about the nature of Jesus Christ. There were those who believed that He was as eternal as the Father and therefore made of the same divine substance.'

'You describe the Holy Trinity.'

'Yes, but an Egyptian presbyter called Arius argued against it. Both views gathered followers and threatened to split the Church. When the argument erupted into violence, the Roman Emperor Constantine called his bishops together for the first-ever Council of the Catholic Church. Arius was defeated and denounced as a heretic. Thus Trinitarianism became the orthodoxy . . . but it was the wrong decision.'

Fatio had turned white. 'Isaac, this is heresy. They can hang you for holding such a belief.'

'Now do you understand why I wanted to leave you out of this?'

Fatio looked crestfallen.

Newton softened his voice. 'I know it's difficult for you to hear, dangerous even, but I think you can carry this burden. Even since that decision, Christianity has been perverted so completely that, even in the Anglican Church, they're unaware of the abomination they perpetrate whenever they accept the sacrament. The blood of Christ and the body of Christ; it's the most evil manifestation of idolatry ever committed. It must be reversed.'

'So, this is the real goal of your work. The overthrow of religion.'

'I work for the glory of God. I am His most humble servant. Fatio, look at me. Do you believe me?'

Fatio looked nervous but he bowed his head.

Newton took him in his arms. 'If I fail, everyone is damned.'

Newton spent the night wrestling with his thoughts. There had been thirteen hundred years of endemic perversion. How many had lived and died under that false doctrine? Every single one of them had lost their souls because of a single decision taken in a desert temple during the fourth century. Even if Newton could begin to turn the Church back now, what was to be done for those already lost? His own mother and father were among them.

Newton felt a tear roll down his cheek. He surrendered to his emotions. The voices in his head spoke their own cacophonous suggestions.

When dawn arrived, he watched the sky with stinging eyes, and a wave of elation overcame him. It was so unexpected that he thought he might weep again. In the grip of the epiphany, he realised there was only one course open to him.

The Philosopher's Stone.

With that in his possession, he could save all humanity – all of those that lived and all of those that had already died. But to do it, he needed to be back in Cambridge with his furnace.

33
Cambridge
1693

Sweat snaked down Newton's back. Hunched as close to the furnace as he could bear to be, he watched the small puddle of gold and lead in the bottom of the crucible. The metals slid together, ready and waiting, approaching the critical point.

Despite the heat, he felt shivery and tense. Days ago he had been stupid enough to admit his maladies to Fatio, thereby giving his companion the excuse to drag him outside for some fresh air. Regardless of their time together, it still escaped Fatio that Newton preferred to persist when experiments failed. The setbacks served to redouble his resolve, not furnish excuses for giving up. Fatio, however, would idle, cursing their lost effort and seeking diversions: meals, turns round the quad, walks along the river. What had happened to the boy's determination these past months? Since their return from High Laver he had been a changed person.

He was sitting now in the corner, listlessly grinding a sample to powder.

'Mercury,' Newton demanded.

Fatio moved the pestle and mortar aside and rooted through the shelves.

Newton watched the noxious potion. The surface would darken at any moment and he would pour in the mercury in a single unbroken flow. He beckoned to Fatio to hurry up.

A fortnight ago they had come so close to success. The sudden fizz and bubble in the crucible had ignited a momentary elation, but it had faded as quickly as the reaction had stalled. Nevertheless, in the wake of the disappointment,

Newton had glimpsed the correct formulation. He saw it not in words or numbers, but in his mind's eye as creatures pacing the primitive earth, grappling each other with their clawed feet, fighting with hooked beaks and spreading their scaly wings to take flight.

He had tried to write it all down, but words were inadequate to describe these processes; they had to be pictured – as in the old alchemical treatises. Newton had clung to the images in his head, bought more gold and taken delivery of more charcoal to prepare the samples again.

For those two weeks he had been unable to sleep for more than a few minutes at a time, lest the beasts escape from his imagination. He had buried himself so deeply in the vision that he now glimpsed them around him in reality. Sometimes it would be a shadow passing the bell tower of Trinity, or a scraping outside his door in the early hours of the morning. Once he heard the bark of a griffin.

Fatio waved a flask in front of him and Newton grasped its glass neck. In the split-second before he slipped the silver metal into the crucible, he checked himself and threw the flask into the darkness.

'No! The purified mercury! It must be sophic,' he shouted over the shattering glass.

Fatio corrected his mistake and Newton tipped the flask. The silvery liquid slipped effortlessly into the crucible, filling it halfway. A moment later the potion began to bubble and the hot breath of alchemy surrounded him. He closed his eyes and he could see the dragon breathing; in its glassy eyes lay the recipe for the Great Stone.

'Look!' he shouted, opening his eyes and wiping sweat from his face.

The froth rose up the side of the vessel so rapidly that Newton thought it would overflow. It was growing in volume. Even in the orange light, the liquid had the unmistakable glow of gold about it.

Fatio was beside him, naked astonishment on his face.

When the first bubble collapsed, Newton thought little of it. Then a second popped and disappeared, followed by a third and a fourth. Within seconds the liquid had returned almost to stillness, filling just half the crucible. A black, scummy top settled over the useless potion. Newton stared at the failed experiment in disbelief.

'Let's get some air,' said Fatio.

'You don't understand what's at stake,' spat Newton.

'You're right, I don't understand. You've had me copying out page after page of recipes and procedures, but I can scarcely understand any of it. What do the green dragon, the doves of Diana and the menstrual flux of Mars have to do with anything? It's all gibberish!'

'How can you not know of these things? We've made each one together these past months.'

Fatio held up his hands. 'I can make no sense of it all. You have me crushing salts and mixing metals, you talk about our final goal and how these things we do are taking us towards it – but I cannot fathom any progress at all.'

Newton stepped away from the furnace, hands pressed into his temples.

'This is no philosophy that I know of,' ranted Fatio. 'I carry the stink of chemicals in my clothes, in my hair. No one will sit next to me at the inn.'

'Then leave! Come and live in college. I'll find you rooms.'

'That's not what I meant.'

'Then what do you mean? Philosophy, gravity, alchemy – all are linked by the active principles. Once we understand those, natural philosophy will be linked from the planets of heaven to the trees of Earth, to the rocks and our bones. We will have a theory of everything.'

'Active principles? A theory of everything?' Fatio spun to face him. 'We would make better progress if we stuck to the second edition of *Principia*.'

'There's no point. Flamsteed will not release his lunar data, so apart from minor corrections there would be nothing to add.'

'My explanation of gravity – I thought we had agreed that it would form a fourth part of the book. Many of my correspondents in Europe are eagerly awaiting it.'

Newton thought his head would split. 'If you truly knew what I am trying to achieve, you would not argue with me. You would work as hard as I do.'

'But we can finish gravity together.'

'No! Until I can understand the motion of the Moon, there is no point in continuing. Besides, this new work is much more important. Once we have the Philosopher's Stone . . . things are just not ready. I have written to Mr Locke, begging him to withdraw the publication of the pamphlet. I must have everything in place before I launch the attack.'

'But we're getting nowhere.'

Newton glared at Fatio. 'You just don't understand.'

'There's no logic or pattern in what you're doing. You're increasingly erratic. You're hiding things from me.' Fatio was becoming tearful. He looked wretched and pleading.

'Listen to me.' Newton leaned against the shelves, beckoning his companion so close that their foreheads touched. He spoke in a whisper. 'Are you familiar with the term *anti-Christ*?'

Fatio tensed. 'I think so.'

'The anti-Christ leads his people into a semblance of religion, one almost indistinguishable from the real thing, except that it's an illusion. It provides no eternal salvation. I believe it was an anti-Christ who perverted the Council of Nicaea. The decision against Arius was not whim or feeble human error, pardonable by God. It was a deliberate act of devilment by Athanasius. I had once thought that the Pope was the anti-Christ, but his evil is nothing compared to the original perversion, introduced so early that it destroyed the true religion within just four centuries of its creation.'

Newton could hear Fatio's shallow breaths.

'Everyone who has died worshipping the Trinity has been damned by the falsity of their belief, by their inability to see into themselves and sense the truth. Even now they're burning – lost to the devil – unless I can succeed with my experiments. This is no time for your childish behaviour, Nicolas. Can you appreciate the urgency now?'

Fatio leaned back, shaking his head.

'I can help them. I can save them.'

'How?'

'The Philosopher's Stone. Remember I told you it would give us power over nature – allow us to control the transformation of matter from one state to another.'

Fatio nodded.

'You've seen the way it can force dead metals to take the form of living trees. It would give us the power of life over death.'

Fatio clasped his hands together. 'Please, Isaac, stop talking. I don't want to hear this.'

'We could resurrect the dead, teach them the true way. We could give them eternal life.'

'But what does that mean for the Book of Revelations, the Apocalypse and the Judgement?'

'It would *be* the Judgement,' said Newton. 'The Church is a crippled stallion, awaiting a miracle. The Philosopher's Stone is that miracle. Whoever discovers it would have the power to grant eternal life.'

'But I've always believed that the Judgement heralded the return of Christ.'

'And with it the reign of the true religion for a millennium. Yes, it does. What you forget is that Christ was a man created by God and graced by divine favour to see the world differently. He then used that knowledge to change the world. But if Christ is not fully divine, then it's possible that another man may be equally blessed with new insight for a new age. That man may be among us already.'

Fatio began to sob. 'You cannot mean . . .'

'My father died before I was born. For all I know, he never existed. Graves are easy to fake. I was born on December twenty-fifth . . .'

'Stop, Isaac. You're sounding deranged.'

'Not deranged: chosen. I have been anointed with God's confidence. Join me.' Newton held out his trembling hand. 'We can lead the world through the Apocalypse.'

Fatio stared at the hand as if it were a burning poker. 'You're insane. Look at yourself, you're a ghost. You think yourself so powerful. Well, I know your secret now, this hideous blasphemy. How far would you fall if word of your Arianism were revealed?'

Newton began to shake.

'You dismiss my gravity hypothesis, when it was you who urged me on. What foul game have you played with me, Isaac? Well, now I can shame you. You think yourself a new prophet, but I'm the one with the power. I can do more than just ruin you, I can see you hanged at Tyburn.'

'You wouldn't . . .' Newton croaked. There was only silence. He couldn't see properly. *Was Fatio gone?* In his place were the foul creatures of his visions. They crept closer, hissing and spitting, beaks open, claws raised. Newton pressed his hands into his eyes, but still he saw them approach.

He screamed out and the darkness took him.

Part III
Force

34
Cripplegate
1703

Halley never understood why his first few days back on dry land were accompanied by the feeling of the waves still rolling beneath him, as if his body could not bear to be separated from the ocean. He had stepped ashore in Deptford a few hours ago, and although he was now crossing the flat grass of Cripplegate meadow, his mind insisted on telling him he was fighting the swell of the Channel. His occasionally misplaced feet and his sway must have given the impression he had been drinking. One stumble almost dislodged the unfamiliar object he was wearing on his head. With a hasty look up to see whether the men scything the long grass ahead had noticed him, he straightened the periwig and continued.

He had bought it an hour ago from a shop near the docks.

'Have you given any thought to style, sir?' asked the shop keeper with exaggerated politeness.

'What about that one?' Halley pointed to a sumptuous creation hanging from its stand like a pelt.

The shopkeeper pulled a face. 'One must be careful to wear an appropriate wig, sir, or one risks being accused of aggrandisement.'

Halley was about to protest at the man's impertinence when he caught a glimpse of himself in the shop's mirror. Roughly shaven, with cheeks ruddy from the wind, the Atlantic salt and the sun, he did not look like a gentleman. His eyes were creased and his thinning hair was knotted into the nape of his neck with a dirty black ribbon.

Nevertheless, Halley spared no effort in explaining his station.

'I do beg your pardon, sir. It is just that you sounded like . . . well, like . . .' The shopkeeper closed his eyes. ' . . . a sailor.'

'I'll take that as a compliment.'

Halley left the shop soon afterwards with the wig in question and headed for the barber.

Now his breath quickened as he left the meadow and rounded the street corner. Before he reached the front door, which he noticed was glossy from a new coating of black paint, William had opened it.

'I saw you coming. Welcome home, sir,' he said with a broad smile.

No longer the boot-boy, William was the manservant of the house. His face had matured, but his eyes still burned with the same willingness to please.

Halley beamed at him but was immediately distracted by three figures rushing into the hallway. Mary ran into his arms. He stroked her hair, tracing the new strands of white. After a few moments he loosened his grip, but did not let her go.

She reached up and touched his new wig.

'Do you like it?' he said uncertainly, easing back to look into her eyes.

'It's curlier than your own,' she said, slipping her fingers into the ringlets.

The other two figures stepped closer.

'Margaret! Katherine!' exclaimed Halley. 'My two daughters are grown more beautiful than I could have imagined.'

That was rather flattering in Margaret's case. She bobbed at the knees. 'Father.'

'Why so serious? Are you not glad to see me returned?' Halley teased.

'Of course I'm pleased.' Her face remained impassive.

Katherine smiled for both of them. Almost a mirror-image of her mother in looks, she squeezed herself into Halley's free shoulder. 'We have so much to tell you, Father.'

How can my younger daughter be fifteen already? Halley thought with a touch of panic.

There was a clatter of footsteps from the landing.

'Careful,' warned Mary as her son barrelled down the stairs.

'Step lively there, sailor,' said Halley, disentangling himself from the women.

'Did you meet any cannibals?' The twelve-year-old jumped from three stairs up, jacket flaring like a bat's wings. Halley lurched backwards as he caught his namesake. There was a collective gasp and Halley felt a cool breeze on his shaven scalp.

Mary was staring at his head. 'I thought you would wear it like a hat, with your hair underneath.'

Katherine was making a poor job of stifling her giggles; even Margaret looked as if she were smirking.

Halley ran a hand over his shaven pate. 'Gentlemen wear it like this to prevent itching.'

'Put it on again, Edmond, quickly,' said his wife.

Halley retrieved the wig from where it lay on the floor and flipped it back on to his head. 'William, keep a watch out for my possessions. They're being delivered from the docks.'

'Certainly, sir. Shall I bring coffee?'

'Yes, please – and some brandy. I think we all need something to steady our nerves.'

Halley crossed the dark landing that night and gingerly entered the master bedroom. Now that the time had come for him to be alone with Mary, he was feeling jittery. All evening she had been quiet, even when Katherine had been making the rest of them laugh with her tales of who was making eyes at whom in church.

When Halley told his tales of storms and dangers, Mary had stared at the rugs or the backs of her hands. At times she had looked at him quizzically, perhaps even sadly, and it had unnerved him.

She must think me grown so ugly. He knew that this voyage had taken its physical toll on him; he had seen that in the wigmaker's mirror – face rounder, jawline flabbier, brow more wrinkled. Yet inside he felt the same as ever; at least, he thought he did.

Mary was sitting at the dressing table, combing her hair by the light of a rush-lamp.

'Katherine looks so like you,' he said from the shadows.

'Flatterer. She's young and pretty. Look at the wrinkles around my mouth and eyes.' Mary leaned towards the looking-glass to examine herself.

'Then my eyesight must be failing me, for I see only the beautiful woman I married. I think I must look like a stranger – an old, unattractive stranger, at that.'

She twisted in her seat. 'Is that what you think?'

'I still love you the same as ever before.'

'Edmond,' she said, 'I need to tell you about Robert.'

A cold weight settled upon him. He could see from her expression what she meant.

'When?' he finally managed to stammer.

'Back in March.'

More than six months ago . . .

'I didn't know how to tell you earlier. You were so happy with the children.'

Halley felt suddenly ridiculous without his hair. He slumped on the bed, his elbows resting on his knees. His stomach had become a great hollow, as if something had been torn from him.

'He'd always pretend not to be interested in my expeditions, yet we always ended up talking about them. Where was he buried?'

'St Helen's on Bishopsgate. I took the children to the funeral.'

'Thank you.' The thought briefly entered his head that he used to return home aflame for Mary, but tonight he just wanted to feel her living warmth.

The overcast sky matched the colour of Hooke's tombstone. It was made of a gritty, rough material, prone to crumbling, and Halley fancied it would not last many English winters. As for St Helen's itself, the rattle of carts on the nearby street and the ever-present echo of the city crowded in on it, making it an unlikely place to rest in peace. *At least you're close to Gresham*, thought Halley, crouching down to brush away a spider. He rested his fingers on the engraving.

Died 3rd March 1703

Halley had been in the warmth of the lower latitudes, somewhere off the Dalmatian coast, taking his magnetic readings and calculating longitudes, watching the Moon melt into the shimmering Adriatic water. And, under the guise of his natural investigations, he had been scouting the coastline at Winslow's behest, looking for German fortifications that could prove useful if troops were ever needed to assault the Papal States. He had not spared a single thought for Hooke or any of the others in the Society.

Are you waiting somewhere, Robert, wondered Halley, *with Grace?*

The cold of the headstone spoke for itself. Halley felt the old doubts rising.

Last night, after the memories of Hooke had assaulted him – the scuttling between coffee-shops, the lugging around of byzantine mechanical contraptions that only Hooke could operate properly, the peevish comments that could be laughed over later – he had watched Mary sleeping beside him. The temporary nature of their lives struck him more forcefully than ever before, and he wondered what he would do if ever Mary were not waiting at home for him.

Footsteps approached.

'Mary said I'd find you here. Welcome back.' Christopher Wren indicated Hooke's grave. 'When did you hear?'

'Mary told me last night.'

'I came round this morning to tell you myself.'

Halley placed his hand on the headstone again. 'He always wanted to die in the spring. Was anyone with him?'

Wren shook his head. 'He sent for Knox, but he arrived too late. Edmond, we've lost others too; Sam's gone.'

'Pepys?'

'I'm afraid so. Back in May . . . and Wallis from Oxford.'

Halley looked up into the heavy grey sky.

'It does mean that Wallis has left the Savilian Chair open,' said Wren.

'The Society will not be the same without them, especially Robert.'

'I'm afraid the Society has not been flourishing this past summer. The physicians have rather taken over, and each meeting is a debate about one grotesque physiology or another. Believe it or not, they discussed the medicinal benefits of cow piss last week. Did you know that if you drink a pint of the stuff, it's sufficient to make you vomit?'

'They actually debated that?'

'And the best time of day to smell flowers.'

Halley looked at the grave. 'Oh Robert, what would you have made of that?'

'We cannot let the Society die,' said Wren.

'Is there a plan?'

Wren turned from the grave and Halley followed. They started down the tree-lined path to the street. 'There are a number of us who wonder whether Newton could be persuaded into the President's chair.'

Halley searched Wren's face. 'Is he fully recovered?'

'Oh yes, leaving Cambridge was the best decision he ever made. Well, that and accepting the job at the Mint. His work there has turned him into a public figure. He's become something of a royal favourite, and I'm told there's talk of a knighthood. So, yes, I'd say he appears himself again, if not more confident than ever, but he's a rare visitor to the Society. Rumour has it that he's finally preparing to publish his book on colour. Why do you look sceptical? You think Newton is a bad choice?'

'Not at all. It's just that I haven't only been at sea all the time I've been away. On my way home I visited Hanover and met Leibniz. There's trouble brewing – big trouble between him and Mr Newton.'

'Tell me more.'

Halley began his story, careful to imply that he had made an innocent jaunt to see Leibniz. In reality he had been working under orders from the Queen to brief her brother, George Ludwig, about the Adriatic reconnoitre.

He had waited with Winslow in a claustrophobic corridor, unable to shake the feeling that he was committing some monstrous treachery by telling a foreign duke about English espionage.

'Just tell him what we've discussed and nothing more,' Winslow had said impatiently, irked by Halley's fretting. 'One day, and probably soon, he'll be the King of England. That's why we've got to start preparing him.'

Halley clasped his hands together. 'It feels uncomfortable.'

'I don't like it any more than you do.' The spymaster jerked a thumb at the dark wooden door, 'He's hardly kingly material; he doesn't even speak English. There are more than fifty others who have a better claim to the throne than he does, but do you know what George Ludwig has going for him? He's a Protestant, and all the others are Catholics.'

He smiled awkwardly. 'And that far down the succession, he'll do anything Parliament asks. So everyone's happy.'

The smile vanished as the door opened. The pair walked in to see a small gathering of finely-dressed men seated around a heavy table. They were silent and expectant, illuminated only by a watery light trickling in from the small leaded windows. Halley guessed from the curve of the far wall that they were in one of the turrets he had seen from the carriage.

The English heir sat at the head of the table. He gestured for Halley to begin.

During the report George bobbed his head as if understanding what his translator was saying, but he looked increasingly nonplussed about what to do with the privileged information. His advisers behaved the same way, nodding with fake sagacity. Halley had seen more convincing puppetry.

The Duke said something official-sounding when Halley concluded his account.

'How do you like it here in Hanover?' inquired the translator with a wan smile.

Halley chose his words carefully. 'It's quite the most serene city I've ever visited. I fear you will find London raucous by comparison.'

The Duke flashed a lopsided smile, throwing an aside to his advisers. The men laughed loudly.

'A chance I'm willing to take. Thanks be to God,' came the translation.

They behaved like men who still could not believe their luck: from a dukedom in this backwater of Saxony, to the monarchy of England on the invitation of Parliament.

'If there is nothing more you require, your Grace, I should like to visit your esteemed librarian,' said Halley.

The Duke's face darkened. 'Leibniz?'

Halley nodded.

'If you must,' said the translator evenly. 'Tell him to hurry up with my family history. He's been working on it for over a decade. Now I'm to be King of England I need it more than ever.'

Halley omitted to recount any of this to Wren, but told him all about his subsequent meeting with Leibniz. His first thought had been that the ducal library could do with being moved to a larger room. The carved and gilded bookcases were overfull, and piles of books buttressed the straining shelves.

Leibniz was perched on the settle in the largest of the bay windows, rocking one leg back and forth. It was adorned with a wedge-shaped shoe and a showy garter tied just below a swollen knee. He was older than Halley had been expecting, plump in the face with dark eyes which were almost lost in small pillows of flesh. His whole attention was focused elsewhere.

There was another person in the recess. The woman mirrored Leibniz's pose and had a figure so shapely it must surely have been turned by a master craftsman on a lathe. Halley drew himself more upright.

'Mr Halley, I am humbled by your presence.' Leibniz's words caught him off guard. 'Allow me to introduce Lady Caroline of Ansbach, soon to be the Duke's daughter-in-law.'

Halley welcomed the opportunity to bow low and collect himself. 'I'm charmed to meet you, Lady Caroline.'

When he looked up again her lips were not answering his smile. Her eyes were as unwavering as the steel grey of her silk dress and transmitted a rather forbidding, critical intelligence.

Unnerved by her appraisal, Halley spoke quickly. 'You must relish your conversations with Mr Leibniz, your ladyship.'

'Who wouldn't jump at the chance to converse with Europe's leading mathematician?'

Halley raised an eyebrow.

'You flatter me too much,' said Leibniz. 'I'm sure Mr Halley knows full well that Mr Newton is also a highly skilled mathematician.'

'Nonsense. If Mr Newton cannot bring himself to publish his new mathematics, then how confident of his work can he truly be?' Caroline rose gracefully from her perch. 'Perhaps I should leave you two to talk.'

Leibniz extended an appropriately deferential invitation for her to stay, allowing her to refuse politely and leave with decorum.

'Mr Halley,' she said with frosty politeness as she left.

'I hope I have not caused Her Ladyship any offence,' said Halley, conscious that if she were marrying the Duke's son, she would one day be England's Queen Consort.

'Think nothing of it. She's young and looks up to me. All European mathematicians now use my notation for the calculus; she admires that.'

'But Newton was first in its derivation, was he not?'

Leibniz's brow knotted. 'I admit that Mr Newton had the first inklings before me – I saw that from the papers he lodged at the Royal Society – but then I developed an independent system, and Monsieur Bernoulli simplified it. That is the way of things. One man takes an initial step, then another carries it forwards, and so on. Mr Newton, for all his skill, cannot lay sole claim to the calculus.'

'You saw Newton's papers?' Halley asked incredulously.

'A few preliminary papers only, back in '72 when I visited London. Your secretary, what was his name? Big man, round face . . .' The German snapped his fingers. 'Collins, that was it. He showed them to me.'

'Does Newton know?'

Leibniz shrugged. 'They were but preliminary pieces, jottings on power series as I recall, but enough to show me that he was thinking in the same direction. You know, I do envy you all in London, coming together every week for discussions. If it weren't for my talks with Lady Caroline, I think I should wither away. She has a keen mind.'

'The Society can be quite rough-and-tumble,' said Halley, still trying to digest the fact that Leibniz had seen Newton's supposedly confidential papers.

'Better that than a vacuum. I have to beg our ambassador in London to send me the Society journal.' He took a breath. 'I once tried to inaugurate a philosophical society here. I designed windmills to drive pumps that would drain the silver mines, and I proposed that the increased wealth thus created should be used to found something in Hanover. No one was interested, in either the windmills or the society. What can one do?'

Halley made sympathetic noises.

'What is Mr Newton really like?' asked the German.

'I find that something of a personal question.'

'Very well, let me come to the point. Lady Caroline and I can admire him for his mathematics but not for his philosophy. His concept of gravity is unworkable.'

'Unworkable?'

Leibniz nodded. 'Nothing happens without sufficient reason, and what reason is there for planets to pull on each other from afar?'

'Perhaps that's just the way the Universe is,' murmured Halley.

'What a lifeless argument – if I may be so bold. The cosmos must be a rational place, so everything must happen for a reason. God cannot have designed it otherwise. Lady Caroline and I worry about the state of English religion, especially as she will one day have to rule there. You and I both come from Protestant countries, yet there seems to be a gulf between us, a laxness in English morals. Tell me, Mr Halley, you and your colleagues do believe in God, don't you?'

'Mr Newton is as devout as they come.'

'Then why do I detect occult qualities in his writing?'

'Never!' But Halley remembered Newton crouched over his furnace, stirring metals.

Leibniz's eyes narrowed. 'Newton imagines gravity as something that can act over great distances without any material intermediary. It is action without contact, a force for no reason. That is an alchemist's view of things – that material objects are threaded like beads upon invisible forces. Has he ever mentioned "active principles" to you? Or "spirits" pervading matter?'

Halley willed himself to give nothing away, even as the passage from the *Principia* danced in his mind's eye: *a most subtle Spirit pervades and lies hidden in all bodies, and by the force of this spirit the particles of bodies mutually attract each other*.

'I see from your expression that he has,' said Leibniz. 'It is irreligious to speak in such a way. A godly Universe is built on overarching principles that philosophers have debated for thousands of years. Newton is no philosopher; he values little observations more highly than whole lifetimes of philosophical thought. His attempt to build a new world-view from the bottom up is flawed, maybe even dangerous. He and those who subscribe to it attack accepted natural philosophy and religion.'

'But his method of investigation defines how we test and prove these ideas. Never before has there been a route to knowledge that allows us to test our ideas. Philosophy is just opinion without experimentation.'

'You cannot possibly travel into the sky and see if he's right about the planets. There are certain realms where only philosophical reasoning can help,' countered Leibniz.

'The observation of Jupiter and Saturn as they draw close in the sky prove that they pull each other slightly off course, then return to their prescribed orbits as they move away from each other. The mathematics is clear; the observation is clear.'

'I have nothing against his mathematics, but it is a tool, it is not a truth.'

'For us, it is the truth. It's the language in which the Universe is written.'

Leibniz tutted. 'That is a dangerous path. Mathematics can assist only when it is paired with philosophy. And philosophy tells us that there is no such thing as action at a distance. It is irrational; it is like an effect without a cause. Newton himself knows that there are errors in the *Principia*.' He began rooting around a desk in the centre of the room and eventually proffered a paper with a flourish.

It was a detailed list of corrections. Halley looked up sharply. 'Where did you get this?'

Leibniz leaned back against the desk, folding one ankle across another. 'I bought Christiaan Huygens's papers for the library here when he died. It was among them. Seems only fair to publish it, don't you think?' Leibniz delicately pulled the paper out of Halley's hand and placed it back on his desk. 'Especially as, from what I hear, Newton's so busy re-coining England that he's turned his back on philosophy.' The tenor of his voice changed, becoming quiet and precise. 'I know of another paper that Mr Newton has written. It's been written anonymously, of course, but one doesn't have to see the lion to recognise the claw marks.' Leibniz watched Halley intently. 'It denies the divinity of Christ, and that's heresy – even in England.'

'I know nothing of this,' said Halley.

'I will allow neither you nor Mr Newton to pervert religion.'

'We search for mathematical truth.'

'At the expense of religious truth. If you were a God-fearing man, you would clearly recognise the subtle poison that Newton spoon-feeds us. You do believe in God, don't you, Mr Halley?' challenged Leibniz.

Halley began scanning the library for a means of escape. He saw that the door was open, but a familiar figure blocked the doorway. There was a look of rapt attention on Winslow's face.

'Do you think Leibniz really does plan to publish that list of corrections to *Principia?*' asked Wren.

'It's already in the press,' said Halley, trying to exorcise the memory of Winslow. *How much had the spymaster heard?* He had not said anything during the evening meal, but Halley thought he had detected a certain overplayed nonchalance.

'Do we hear anything of Mr Fatio and his theories these days?' asked Halley.

'We hardly see him, and never on the rare occasion Mr Newton attends.'

An unexpected laugh burst from Halley. 'Remember! Robert used to call him "Newton's ape", and the "perpetual motion man". "Here comes the perpetual motion man", he'd say.' Halley could still picture the dainty approach of Fatio that the phrase evoked.

Wren began to chuckle. 'He used to sign them in together on the same line of the attendance register, as if they were a single entity: *MrNewtonFatio*.'

'Do you think they were together? You know . . .'

Wren regarded Halley closely. 'No one knows – and best to leave it that way, with the law the way it is.'

'So, has anybody asked Newton about assuming the Presidency?'

Wren looked over at Halley, suddenly bashful. 'We were rather hoping you'd do that.'

Halley heard the percussion of the Mint before he saw it. The thumping of the mechanical cutters lifted through the London air to echo around the Tower. The Mint itself was a collection of ramshackle wooden huts strung along the outer of the fortress's walls, and Halley was led under close escort to Newton's office.

'I see we are both wigged these days,' said Newton in lieu of a greeting.

'And a confounded nuisance it is, too.'

Newton peeled off his wig and dropped the white hairpiece to the floor. 'Let us at least be comfortable.'

Halley relaxed and removed his own wig. 'How are you?'

'I think we've known each other long enough to dispense with vague pleasantries. What can I do for you, Mr Halley?' As an afterthought Newton indicated a chair for his guest.

'In November the Royal Society will be looking to elect a new President. The Fellows wondered whether you would consider the position,' said Halley.

Newton stood up immediately.

'I confess I thought you had quit philosophy for good,' added Halley, quickly standing again.

'So did I,' said Newton, turning to gaze out of the windows at the other huts. 'So did I . . . It seems that every time you turn up unannounced, Mr Halley, you have a diverting question for me. Would I be free to reorganise the Society?'

'Completely.'

'No allies of Hooke ready to lay siege to me?'

'Robert has gone for good,' said Halley, irked by the question. He concentrated on the older man's ink-stained fingers, watched them flexing.

'From the few visits I have made to the Society, I should say that natural philosophy has lost its way,' said Newton at last. 'I should like to turn it back. I've been thinking of publishing my theory of colours.'

'I'd heard a rumour,' said Halley.

'The book is written and ready for the printers. I have no need of your assistance this time; these days there are plenty who wish to help.' Newton returned to his chair. Halley did the same. Newton then spoke in a curious, hurried voice. 'When I was first upon the colours, my design was to continue my investigations until I understood how the light rays themselves were produced. By that I mean how crude matter becomes bright. But as you know, I was . . . discouraged in this endeavour by a certain lately departed individual. Now I find I have neither the time left nor the energy to set upon these investigations myself. So, I've listed the questions I wished to answer in my book – I'm calling it *Opticks*, incidentally. Instead of a conclusion I give sixteen queries, each one answerable by experiment. It's a manifesto for natural philosophy and the experimental method.'

'Then the Presidency would suit you?'

Newton ignored him. 'I do have one question you can help me answer immediately. I hear that philosophers abroad are crediting Mr Leibniz with the invention of the fluxions. If I include some of my early papers in the *Opticks*, do you think that would settle the issue? I have no desire to become dragged into a dispute with him, but I will have my priority in the ideas acknowledged.'

Unwilling to provoke an explosion, Halley decided not to mention his troubling visit to Leibniz. 'I can think of nothing better than to clarify the matter. You've been too modest thus far.'

Newton nodded. 'There is one more matter that nags at me. I'd also like to publish a new version of *Principia* with a complete theory of the Moon. I've tried to put it aside but it concerns me still.'

'I cannot tell you how important an accurate theory of the Moon would be. I tried to calculate longitudes from lunar observations while I was at sea, but they were dozens of miles off. With the night close around you or a storm blowing, a dozen miles is enough to have you on the rocks in the blink of an eye.'

'Then that must be a priority for us, do you not agree?'

'I do. It's a frightening thing to be at sea and in nothing but the hands of God.'

'Surely there is no safer place to be?'

Halley's frowned. 'You know what I mean. But Flamsteed is stubborn with his observations.'

'He must be hiding something.' Newton drummed his fingers on the desk. 'The Royal Society must be returned to its guiding principles: Galileo's experimental method. And Flamsteed must be brought to heel. I will need a team of experimenters. Three will suffice; paid, of course, so that they will follow orders. The Society must progress. Gresham College is not highly enough regarded. We must begin the reorganisation at once.'

'So, you accept?'

Newton brought his fist down, setting the desk and papers quivering. 'I will. And Mr Flamsteed and the theory of the Moon will be my first projcct.'

35
London

It was a week later when Halley met Wren in a packed Garraway's to celebrate Newton's acceptance. Not long into their conversation, the subject of the vacant Savilian professorship at Oxford came up again.

'You'd be a strong contender,' urged Wren over the hubbub. 'You're an Oxford man yourself.'

'I ran away from Oxford to survey the southern stars. They only granted my degree at the King's command.'

'Really? I didn't know that.'

'You did at the time.'

Wren puffed out his cheeks. 'Well, they don't remember it either.'

'How do you know that? Have you been speaking out of turn?'

'Perish the thought. They approached me, asked me to invite you.' Wren fixed him with a look. 'The professorship is yours. All you have to do is say yes.'

When Halley arrived home, it was hard to tell where the bodies finished and the material started. Mary and the girls were inextricably woven into rich folds of shot silk that he had bought for them in Gibraltar.

'I have an important announcement,' Halley said in a voice that sounded more pompous than he intended.

'You're with child?' asked Katherine.

'Katherine! We'll have none of your cheek here,' said Mary.

'That's how you told us young Edmond was on the way.'

Margaret shushed her.

'Thank you, ladies,' said their father. 'I have been offered the Savilian professorship of astronomy in Oxford. I intend to take it.'

'Oxford?' gasped Mary.

Halley nodded. 'I've been so busy helping others that my own astronomy is all but forgotten. I want to return to my work on the comets, plotting their orbits.'

'But between Oxford and sailing we'll never see you.'

'No, you misunderstand. I've finished sailing.' He hurried to the chair next to Mary's at the table, diving into the fabric ocean himself.

'But you were talking about returning to the Atlantic only last week. I heard you the night the Bowers came for supper.'

'I know that's what I said, but I've changed my mind. It's just–' He checked himself. He did not wish to reveal how he lay awake at night, watching her sleep and grieving for her even though she was still alive. He should have been mourning Robert and Samuel, but somehow his mind had transferred it all into the terror of losing Mary. He counted the new lines on her face and the new threads of silver in her hair, and regretted not seeing every single one form. Time was stealing away. 'It's just that I miss you and the children more these days. I didn't realise that until now.'

She studied him and he could see in her eyes that she had read his mind. He was certain of it when she squeezed his hand tightly under the fabric.

'Oxford it is, then,' she said.

'We would not need to move lock, stock and barrel to Oxford. There's a house that comes with the position, but we will retain this one and divide our time between the two cities. After all, Newton is going to change things here at the Society, and I'd like to be a part of that. In fact,' said Halley, glancing at the tall clock in the corner, 'I think things may be changing as we speak.'

There were several tall ships at anchor, rising like spectres from the water, masts as bare as the autumnal trees. Up in the curve of the river, where the Thames grew deep and dark, a three-masted Navy pink was unfurling her sails to begin the journey towards the open sea. Standing on the wharf, Newton shivered despite his heavy coat.

He picked his way around the tea-chests and sailors, cut between the warehouses and began a steady ascent of Greenwich Hill. The observatory seemed to draw no closer, and soon he was shaken by a coughing fit. He marched on. The exertion darkened his mood further.

By the time he reached the shadow of the observatory wall he was breathing heavily. He leaned against the brickwork, having no desire to meet Flamsteed while panting like a dog. Pushing away from the wall, he came face to face with the Astronomer Royal.

'I didn't mean to startle you,' said Flamsteed. 'I saw you from the window, but when you didn't appear I came to make sure you'd survived the hill. It's quite a climb.'

'I'm perfectly fine, thank you,' retorted Newton.

The two men laboured up the steps into a small hallway, which led to the cramped sitting-room. Despite the chill, no fire had been lit. They watched as a stout woman waddled in carrying a tray of coffee.

'I don't believe you've met my wife. Margaret, this is Isaac Newton.'

Newton acknowledged her but did not look up. He would have preferred to meet Flamsteed alone.

'I've heard all about you, Professor Newton,' she said in a wheezy voice.

'I'm no longer a professor. I left Cambridge almost a decade ago.'

Margaret thrust the tray between the two men. 'Begging your pardon, I'm sure.' She inclined her head and turned on her heel.

Flamsteed looked vaguely amused by his wife's bad manners.

'Can we expect you at meetings now?' asked Newton abruptly.

'Whenever I can get there, but my joints make travelling painful these days.'

Newton sipped the coffee, already bored with Flamsteed's joints. The drink was gritty and lukewarm. He replaced the dish on the table and wiped his fingers down the sides of his mouth. 'Do you remember coming to my lodgings in London some years ago?'

A flicker crossed Flamsteed's fat face. 'Yes.'

'You left a biblical passage for me to read.'

'I did.'

'A warning against false idols.'

Flamsteed did not speak.

'Whom were you referring to?' asked Newton.

'Isn't it obvious?'

'No,' said Newton through tight lips.

Flamsteed waited a number of maddening moments. 'Edmond Halley.'

'Halley!' Newton wanted to set the fleshy face before him shaking with the flat of his hand. To think he had assumed Flamsteed about to confess to Arianism!

'Where does one start with that man? He's arrogant, intemperate. He's an adulterer: cuckolded Hevelius all those years ago in Danzig.'

Newton glared at Flamsteed. 'Mr Halley's personal affairs are his business. It's his astronomy I'm interested in.'

'Pah! Slapdash and prone to error.'

'Whereas you're a perfectionist.'

'The royal star chart demands nothing less, wouldn't you agree?'

'Quite so. But thirty years is a long time to be working on it, don't you think?'

'Twenty-eight, Mr Newton. Tycho Brahe was not harried. He collected observations of just a thousand stars over twenty-five years, with dozens of assistants. Kepler then took another twenty-seven years to calculate and produce the tables. I've measured more than three thousand stars repeatedly and performed or supervised the calculations myself. As star catalogues go, I'm working rather quickly.'

'Hmmm, but how long before you publish your results?'

'They're all but ready.'

'Ready,' said Newton, attempting to sound casual.

Flamsteed nodded.

'You must be planning a lavish presentation to do justice to such a monumental work.'

'Of course.'

'An expensive presentation?'

'That is my intention.'

'I think the Crown should support the printing, don't you? You're the Astronomer Royal, after all,' Newton purred.

'Do you really think they would pay for publication?'

'Prince George takes a great interest in natural philosophy.'

'But how would I raise such a matter with tact? I do not know Their Majesties well.'

Newton smiled. 'I could do that for you. Haven't you heard? I'm to be knighted. I have the royal ear, and these things are always better coming from supporters. Shall I raise it with Her Majesty, to save you the embarrassment?'

Flamsteed nodded dumbly.

A steely exultation washed over Newton. He could almost see the data in front of him.

Hooke's apartment had rotted along with its owner. Neglected in the months since his death, there was a palpable decay in the air, and not just from the rancid food in the kitchen. The stench was so bad in places that Newton had to raise a handkerchief to his nose – not that it made any difference.

Newton stalked the depressing chambers, still tingling from Flamsteed's acceptance. The publication of the Greenwich star catalogue would be the first triumph of his reign as President, shortly followed by the new edition of the *Principia*, and then *Opticks*, with its queries for investigation. Within a year, he thought, he would change navigation for ever, and set the Royal Society on the path to completing his work on light. If enough experimenters could be brought to bear, all his queries would be solved in what remained of his lifetime.

In the basement he found workers emptying the display cases, tucking the artefacts into straw-lined boxes. 'Wait,' he called, and lifted an object from a workman's grasp. It was the little desktop telescope he had fashioned and presented to the Society thirty years earlier. Tarnished now, and all but useless, it had been left to perish by Hooke. Newton passed the telescope back for packing and renewed his prowl. Occasionally he diverted himself with a contraption or a book or a manuscript, mostly he just scowled. *So much junk.*

'Whatever you don't take will be burned,' said a bailiff.

'Burned,' echoed his companion.

Passing back through the familiar living-room where the Society used to meet, Newton made his way to the stinking bedroom where Hooke had died. He cast a disdainful eye over everything, fighting the desire to retch at the stained mattress – the sheets were already on the fire outside. Surely the rest of the bed must join them soon.

He stared at Hooke's microscope. Positioned on a table at the end of the bed, it was probably the last thing that the Gresham Professor ever saw. A dark urge stirred within Newton. He touched the etched leather of the main tube and the turned wood of the eyepiece. His finger left a clean streak on the dusty object. He grasped the tarnished brass support and lifted it up.

He thought of Woolsthorpe and the plague summer, when he had buried himself in Hooke's book *Micrographia*. Perched on the windowsill in his bedroom, overlooking the orchard, he had opened up the pages and marvelled at Hooke's drawings of gnats and fleas and other minute natural objects. All of them were rendered in exquisite detail because of the instrument he now dangled above the floor.

Newton had been full of admiration for Hooke and desperate to win his approval. He had stayed away from the Royal Society until he had achieved something big enough to impress with: a theory of colours, proven with a crucial experiment. Yet look how that had turned out.

Newton's fingers loosened on the microscope.

He relived the hot confusion that Hooke's assault had provoked, recalling the effort not to scream with rage, and the subsequent sitting into the night attempting to reconcile the insights of the *Micrographia* with the disingenuous missive that had carried the attack. And now Hooke's most precious instrument was moments from destruction. Newton could already picture it in fragments on the floor. He lifted his arm.

But he could not let it go.

With a grunt he turned from the room and carried it down the staircase.

'I'm finished here. Keep this safe.' He handed the microscope to one of the bailiffs, who placed it near a pile of other things, including the portrait of Hooke that used to hang over the fireplace.

Newton stared at the image. The Gresham Professor was seated and gowned, head cocked at an inquisitive angle, his curved spine disguised.

Hardly an honest portrayal, Newton thought.

'I'll take care of this,' he said.

The painting was light enough for him to carry under his arm. He swept from the room.

In the quadrangle outside the fire was crackling and giving out a fearsome heat. Two men with pitchforks were stoking it with Hooke's unwanted possessions. One stopped to mop his brow with his forearm and eyed Newton with curiosity.

The president ignored him and stepped closer, turning his face away from the flames. He swung his arm and sent the canvas curving through the air. The oil paint caught the flames with a hiss.

Where once there had been pink silk filling the front room, now it was straw and tea-chests. Margaret had set herself up as foreman and Katherine was proving a reluctant worker. Even so, one by one the family possessions disappeared for the move to Oxford.

'We don't need to take the whole house,' said Halley plaintively. 'We'll still be here as much as there.'

'We're not taking everything, just a few essentials,' said Mary.

Halley rolled his eyes and almost cried out.

Through the window a familiar black carriage was visible, parked on the opposite side of the road.

He slunk to the front door and rushed across the road. The carriage door was already open.

'What are you doing here?' he asked.

'Thought it was time I paid you a visit at home. You have such a hectic social life, but you never seem to invite me.'

Halley pointed his finger at Winslow's face. 'Don't involve my family. They know nothing of you and that is how I like it.'

The only thing preventing the astronomer from seizing the scraggy neck and making those bulbous eyes bulge even more was the dark shadow of a henchman within the carriage.

'But Edmond, I want to talk to you,' said Winslow mockingly.

'I've been working.'

'What on?'

'Comet orbits –'

Winslow held up his hand. 'Not interested.'

'Look, I'm finished with sailing. What further use can I be?'

'You're going to Oxford, for a start. Plenty of Jacobites there to interest me. They're very upset that the Hanoverians are going to rule, you know.'

Halley let out a derisive sound. 'Is there anywhere you cannot unearth a conspiracy?'

Winslow's mouth creased into a lopsided grin. 'It's my job.'

'Well, it's not mine.'

'I want to talk, that's all,' Winslow whined in his ear. 'I'm interested in Newton.'

So he had *been eavesdropping in the library in Hanover.*

'Now, shall we discuss this in the street, or in private? Step inside. They let me choose my own carriage now. A job's got to have some perquisites, don't you think?'

Halley folded his arms and stayed his ground.

'What's Newton's agenda?'

'No agenda, save understanding God's laws.'

'It's interesting you should mention the Almighty.'

Halley sighed. 'If you have something to say, say it.'

Winslow studied him for a moment. 'The Hanoverians are devout Protestants. They've been promised a pure, God-fearing England of Protestant design. I will ensure they get it.'

Halley feigned indifference. 'I fail to see what that–' A pain lanced his wrist. Winslow's bony fingers were wrapped around it, digging into the sinews on its soft inner side. Halley unfolded his arms but Winslow's grip stayed firm.

'Don't play innocent with me. We both know exactly what was said in Hanover. I'm growing rather suspicious of this so-called Royal Society. Newton to be knighted, for goodness's sake. What is becoming of the peerage? Springs for carriages to save our derrières, charts for navigation, clocks for timekeeping – I can see the value in all of that,' Winslow said. 'But spirits passing through space, changing our world and moving planets – what possible use has mankind for that knowledge? Except to undermine God, and by doing that you undermine all earthly authority based on God.'

'That's not our aim. We're simply curious men. Newton's about to publish another book. Written in English: plain language. He seeks only to prove God's existence by observation.'

'A book for everyone to read? Now that does concern me.' Winslow tightened his grip.

Halley yelped. Specks of blood formed around the tips of his captor's fingernails.

'Why do you need proof of God? Isn't faith enough? Start questioning divine wisdom, and what's to stop you questioning earthly authority? Let me put it to you another way, Mr Halley: the monarch is the head of the Church. Any attack on religion could therefore be seen as an attack on the state, and an attack on the state could be seen as treason.'

'Ridiculous.'

'Is it really?' Winslow's eyes burned. 'Look around you next time you take a walk. All those people going about their daily lives, trying to make enough to eat, to protect themselves in their old age. What stops them just taking what they want from each other? There are plenty of dark streets where murder is easy. What makes society work? What makes people better than the beasts in the forest, eh?' He jerked Halley's hand, eliciting another cry of pain.

'Common decency.'

'Common decency? What dream world do you live in? Divine retribution, Edmond. The Day of Judgement. Without the threat of God hanging over them all, London would be in chaos. And, trust me, it's bad enough already. Power is changing in England. It used to lie solely in the Church, but since Henry Tudor sacked the monasteries it's all changed. Power went to the monarch, but these past few decades it's been unseated again. Now it's coming to roost in Parliament. And to keep order, they need a religious population, not a new way of looking at nature.' Winslow released Halley's wrist with a sound of disgust.

The astronomer cradled his arm. The throbbing red nail-marks looked like a succession of crescent moons.

'What troubles me, Edmond, is that if Newton or you explain everything, what room does that leave for God? Ever thought about that?'

Halley looked up sharply. He could picture Hooke's incredulous face that night up on the Gresham observing platform long ago, when he had asked him that exact question.

A knowing look spread across Winslow's face. 'Newton may seem invulnerable now,' he said, leaning out to reach for the door, 'but all that could change when George Ludwig arrives in town. Think about it. Don't get caught on the wrong side. I'd hate that to become a family trait.'

'Are you referring to my father?' Halley stammered.

'I was at the Tower that day. You know, the older you get, the more you look like him. Especially, when you're scared.' Winslow winked and slammed the door.

Halley watched the black vehicle trundle away. Legs unsteady, he felt as if he had just been robbed in broad daylight.

36
Cambridge

The Cambridge streets were lined with academics, their woollen and silk gowns billowing in the breeze. The dons' presence and the bellowing of the town crier had drawn the townsfolk out of the taverns to swell the ranks, and everyone was watching the trotting Horse Guards and the carriages of the Royal Procession. Newton looked out from his vehicle in the line and watched the faces in the crowd.

It had crossed his mind to wear his old gown, but when he had looked at it in the wardrobe that morning he had shuddered. He had opted instead for a jacket of the brightest bottle-green, trimmed in thick black braid, with matching waistcoat and a velvet tricorn for his head.

He had selected a new wig, four inches longer than his last. It was only a fraction shorter than Prince George's, a fact that had raised a few eyebrows at the palace that morning. Waiting for the carriages to arrive, one of the royal attendants had discreetly approached him with a pair of shears on a cushion and a hastily-appointed barber. Newton had shooed both away.

Faces flashed by. Each one conveyed a mixture of fascination, jealousy and cynicism, and served to distract him from the approach of his destination. It had certainly not been his idea to conduct the investiture at Cambridge. He had hoped never to set eyes on the place again.

Those final months in Trinity still haunted him. He could recall the heat of the furnace and the sting of the chemicals. As much as he hated to admit it, he could still sense Fatio's absence from his side. He had never seen him again. Months later, he had heard that the boy had left the country; then a paper appeared, written by Fatio, attacking Leibniz for claiming the invention of the calculus. Most recently the gossip had it that the Swiss mathematician was back, teaching in Spittlefields again. Newton's urge to see him had increased tenfold.

The carriage drew to a halt. Ahead, footmen jumped from the back of the Queen's carriage and opened the door for her. The university hierarchy jostled, greeting her and her husband with supercilious smiles and too much bowing.

Idiots.

Newton took great pleasure in being announced to the College Master by the palace spokesman as if he were a stranger.

'Would you like to see your old room?' the Master asked, beaming at him.

'No, thank you.' Newton strode on to rejoin the royal party.

The college dining-hall was smaller than Newton remembered. He made his way through the gloomy ranks, past the portraits where the windows should have been and up to the top table. He was shown to his seat, next to the Queen's, and stood watching the royal couple make their way through the rank and file of the Fellowship, crowded at the trestles stretching the length of the hall.

Queen Anne was unusually homely for a monarch. Her rosy cheeks and round face reminded Newton more of a cook, and he imagined her bustling round a kitchen rather than a throne-room. Only her drooping expression gave away her melancholy, a nagging doubt that God was displeased with her.

Newton had heard the details discussed often by courtiers: thirteen miscarriages and stillbirths, four dead infants, one dead child. Her inability to produce an heir had led to the crisis that had resulted in the House of Hanover's good fortune.

Newton concentrated on his rehearsed small-talk during dinner. She wheezed between mouthfuls. With every deep breath her ample bosom swelled. Newton looked away.

'You don't eat very much,' she observed.

'I find myself replete with the honour of the occasion,' he bluffed.

Her face glowed with pride. 'Then I have finished, too.' She stood, sending the room into a clatter of hastily replaced cutlery and scraping chair-legs, and walked to the front of the table, where she surveyed the lines of diners.

'We are gathered today to honour a man of extreme intellect and moral conduct. Many men have served this country with greatness and distinction, but few have contributed so much in more than one field . . .'

Newton watched the crowd as she spoke.

'Step forward,' commanded the Queen, breaking Newton out of his surveillance. 'Either one of your achievements, here in Trinity College or at the Mint, would have been enough to see you knighted.'

A page handed her the ceremonial sword and backed away. Newton knelt before his monarch and dipped his head. He sensed the weight of the weapon on his shoulders, imagining the metal's icy touch even through the thick padding of his jacket.

They say that the executioner touches the axe blade upon your neck before raising it for the fatal blow . . .

'I now pronounce you Sir Isaac Newton.'

As the procession wound its way to the chapel, Newton hung back to fall in step with the Queen's Consort, Prince George. As he drew close he saw that the man's jacket was splattered with remnants of his dinner. On the few occasions they had met before, Newton had talked and George had snorted his way through the conversation.

'How are the old experiments coming along?' the Prince asked breezily.

'Very well, Your Royal Highness,' said Newton. 'The Royal Society makes excellent progress.'

'Good, good.'

'There is one area, however, about which I am concerned.'

'Oh dear,' George guffawed as if Newton was joking.

'It's a most lamentable situation, sir.'

'I'll wager . . . Oh, birds!' George pointed at a flock of black birds taking to the wing.

'It's Mr Flamsteed at Greenwich. I hear that he seeks royal patronage to publish his star catalogue.'

'Does he now? Well, I cannot say I blame him, it is my observation-ory, after all.' The Prince nudged Newton with an elbow.

'Quite, sir, but despite Mr Flamsteed's repeated assurances to me that the work is almost complete, he never manages to present his manuscript.'

'What prevents him?'

'He may not be a man of true learning like you and me,' greased Newton. 'I fear that he may have something to hide, some error in his observations or calculations that he has made. But I cannot ascertain whether it is laziness or incompetence.'

'Dear me.'

'You know, it has been on my mind for some time, sir, to invite you into the Royal Society. We have languished too long without a royal Fellow, and I know of your fondness for the mechanical arts.'

George stopped dead in his tracks. 'You mean it?'

'Of course.'

The Prince's eyes widened like a puppy's. Newton thought for an awful moment that he might start bouncing on the spot. 'Then I jolly well will gravitate along to the next meeting. I'm full of ideas, you know.'

Newton winced, and hoped he had disguised it. 'Your noble duty would be fulfilled in the first place by one simple thing: a decree for the Royal Society to appoint a supervising committee to the observatory in Greenwich. We could then examine Mr Flamsteed's work to determine whether it is fit for patronage.'

'Of course, of course. You must do whatever you can. See my office when we're back in London.' George turned to one of his aides. 'This man will be coming to see you. He wants to oversee the observation-arium-ory – what's it called?'

'Observatory, Your Highness,' supplied Newton.

'Yes, that! At Greenwich. I want you to see to it that he can.'

The aide nodded and Newton wanted to shout in triumph.

Flamsteed, I have you!

The Astronomer Royal read the piece of paper with an open mouth.

'The trickster! I won't comply.' He thrust the inked sheet back at Halley, who refused to take it.

'It's a royal warrant,' he said. 'You have no choice. Aren't you going to invite me in?'

Flamsteed filled the doorway. 'Newton offered to help secure patronage to pay for the costs of publication. Nothing was said about his assessing the data first.'

'You cannot be surprised that some level of review is necess–'

'My data are my own, and I will publish when I please.' Flamsteed let the warrant drop. The breeze caught it and sent it skating across the ground. After tangling briefly round the legs of the bay mare that had drawn Halley's carriage, the warrant fluttered away. Halley thought about going after it but decided to stand his ground.

Flamsteed studied him from head to toe. 'You used to be such a fine young man, Edmond.'

'This is not about me. It's about you doing your duty.'

'Duty!' Spittle flew from the crabbed mouth. 'I've been the Astronomer Royal since the fourth of March, 1675. Every clear night – no matter how bitter – I have observed. I've observed until my body is broken, my bones are weakened and arthritic.' He thrust the ruined fingers of his hands towards Halley. 'I will not be lectured by a libertine.'

Halley stared at the bloodshot eyes. 'Mr Newton needs your data to complete a theory of the Moon and make our navigators safe at sea. It's the very reason Greenwich was established in the first place.'

'You deceive yourself. Even if Newton concocts his magic theory, how is a navigator going to use the information? It will require precise observations to be made, and how are you going to achieve those from the rolling deck of a ship when you cannot keep a telescope pointed anywhere? You must know that from your own sea voyages.'

'Some way will be found,' said Halley stiffly. 'Now, John, your data, please? I'm here to collect it.'

Flamsteed shook his head in disappointment. 'I should have known you two were in league.'

'We must all work together. For too long we have pursued our own goals without focus. Newton's physics gives us a priority, a way of thinking. We are . . . we are Newtonians now. Anything else smacks of mediaevalism and superstition.'

'Does it? After the visitor I had here the other day, I think it would be wise to distance myself from you altogether.'

Flamsteed's words caught Halley. 'Who came to see you?'

'It appears your new philosophy may not be quite as acceptable as you think.'

'Who came to see you?'

Flamsteed said nothing.

'Very well.' It was all Halley could do to control himself. 'Where are your observations? In the name of the Queen I demand logs, ledgers, reductions, everything.'

'I won't comply.'

'Face it, John. You're an old man. I can take whatever I like.' Halley climbed the final step. His elbow pressed into the doughy torso as he barged past the old astronomer.

Newton's stomach was churning as he stepped into the rowdy gin-house. He had hoped never to see one of these interiors again. This particular example reminded him of a cattle pen on market day, with the patrons lurching around like beasts.

He had been forced to frequent such drinking establishments in his early days at the Mint, feigning friendships, buying drinks and loosening tongues in his efforts to track down counterfeiters. Through eavesdropping on the conversations around him and asking the occasional seemingly-innocent question he had compiled his evidence: the names of the ringleaders, the places they worked, their rivals and allies. He had logged every scrap of intelligence until he knew more about counterfeiting than any individual miscreant involved in the process.

Then he had smashed them.

Compared to those investigations, tracking down a Swiss mathematics teacher had been simple.

Fatio was draining his glass when Newton arrived at the table. The young man looked bedraggled, even though the day was dry. His ears protruded through lank tresses and his once pert cheeks had sagged.

'Isaac.'

'You've changed,' said Newton, wondering whether this had been such a good idea. Where was the beautiful boy he remembered?

'How did you find me?'

'I heard you were back in England.'

'Please, join me,' said Fatio, wiping his mouth.

Newton drew up a chair and sat opposite. Feeling suddenly awkward, he thrust the book he was carrying into Fatio's hand. 'I've brought you something. It's my *Opticks*. I thought you'd like a copy.'

Fatio opened the stiff leather binding and flipped through the pages, pausing here and there to skim a passage or trace the lines of a diagram with his eyes. Newton had quite forgotten this hasty, careless facet of the man.

'You publish your fluxions at last,' said Fatio, reaching the back of the book.

'A few early papers, that's all. I tire of Mr Leibniz constantly snapping at my ankles.'

'I published in defence of you against him a few years ago.' Fatio spoke falteringly, as if he was unsure whether it had been a wise thing to do or to admit. His accent had thickened, but that might have been the gin.

Newton said, 'Let me see if I can remember the passage: "I recognise that Newton was the first and by many years the most senior inventor of calculus . . . As to whether Leibniz, its second inventor, borrowed anything from him, I prefer to let those judge who have seen Newton's letters and other manuscripts, not myself." I was – am – sincerely grateful for your support.'

Fatio shrugged. 'I thought perhaps you might have written to me when you read it.'

Newton looked at the stained tabletop to avoid answering. He had still been angry at that stage, and had considered the document a rather pathetic attempt to curry favour.

'I've followed your career at the Mint,' said Fatio.

'It was a good move, especially after . . .'

'After *us*?'

Newton gave a small nod, but then added, 'After my temporary insanity.'

Fatio looked forlornly into his empty glass. 'What does it feel like to send a man to his death? The clippers and the counterfeiters, I mean. You pursued them without mercy.'

'I did my job,' said Newton. 'But I never attended the executions.'

'I did. I wanted to see the men you had beaten. Did you hear how Challoner begged for his life up there? Swore to the crowd that he was being murdered.'

'A lie is a poor way to meet one's Maker.'

'He'd been dragged through the street on a sled. He was covered in mud and filth, shivering with cold. Stone cold sober. But he climbed the ladder and put on the hood before presenting his head to the noose.'

Newton made no response. *Where was this leading? This whole encounter had been a dreadful mistake. The past was the past.*

'I miss you, Isaac,' said Fatio quietly.

'I miss you, too,' Newton found himself replying.

'We're both changed men,' said Fatio.

'We are.'

'But I've been thinking a lot about our final conversation. I was too young. I want to–'

Newton held up a hand. 'Not here.'

'Here's as safe as anywhere to talk about it,' said Fatio.

Newton almost laughed. 'That's what the counterfeiters thought.'

There was a manuscript on Fatio's desk that the younger man clumsily hid, but Newton had already recognised one of the diagrams; Fatio was still working on his impossible theory of gravity.

The mathematics teacher was lodging in a room only half the size of the one Newton remembered. The plaster was damp and a used chamber-pot sat in the corner. Fatio made a bad job of emptying it through the window, then clattered the pot back on to the floor.

'You wanted to tell me something,' prompted Newton.

Fatio nodded and sat on the bed.

Newton joined him, careful not to position himself too close. 'There are some things, Nicolas, that are perhaps best forgotten.'

'No. I was naive before. You scared me. But I've come to understand what you said and recognise the truth. Tell me, do you still have the pamphlet you wrote for John Locke?'

'It was destroyed,' Newton lied. He had no idea where it was; floating around on the continent, he assumed. Thankfully, he had not signed it.

Fatio grimaced. 'Too bad. Your position in society now would be . . .'

'You must never tell anyone what I said. It was a symptom of my delirium.'

'But you were right.' Fatio looked at him squarely. 'Absolutely right. I see it now. I've been reading Revelations – I cannot believe that I was so blind to it before. I want everyone to know.' With a flurry of movement Fatio dropped to his knees and bowed his head. 'Bless me, Isaac. I cannot tell you how much it would mean to me. Please, bless me.'

'Nicolas, you're drunk. Get up.'

'No,' pleaded Fatio, 'the Saviour is referred to as "the Lamb". You were born at Woolsthorpe, your estate, your flocks of sheep . . . It all fits. And everything you said before, I remember it all. It all fits. I want everyone to know.'

Newton reached out and touched the man's greasy head to quieten him.

'Let me repair the damage between us,' begged Fatio.

'There's nothing to repair on my account.' Newton removed his hand.

'Truly?' Fatio lifted his head; his eyes were filled with tears.

'Truly.' Newton forced himself to speak but his mind raged. *What a disaster this was proving to be. Fatio was clearly insane. What would happen if he made these ludicrous beliefs public?*

His confusion mutated into cold logic. He could not ignore this. He had to protect himself.

He looked at thc babbling man.

The threat had to be removed. For good.

37
Hanover, Saxony

The library's windows were dark and shuttered. The rain beat against the hidden glass and a single candle flickered in the draughts. Leibniz sat at his table, working in the restless flame's pool of light.

'Another book for you, Mr Leibniz.'

Leibniz could hear the disdain in the servant's voice. The family history was still not complete. As a result the Duke was refusing to see Leibniz about anything other than business directly related to the book. The snub was damaging the philosopher's reputation throughout the household.

So be it. He could survive without the servants' respect. All that mattered was that he still saw Lady Caroline almost daily. She was proving a formidable scholar. Having just completed Thomas Aquinas's thirteenth-century work *Summa Theologica*, she could now debate theology as well as any man, if not better.

Leibniz accepted the book from the servant with feigned indifference. It was Newton's new book on colours. Looking at the title page, his mouth went dry at the mention of addenda: *Also two treatises of the species and magnitude of curvilinear figures.*

Why was Newton publishing his calculus papers now?

It took just seconds to realise that these were old works, dating to Newton's conception of the technique.

What was going on in London?

First, there had been Fatio's pathetic accusation of plagiarism. It was well known that the man was an intimate of Newton's, so it could hardly have been published without his knowledge. Now, here was a new book with a couple of outdated papers tucked in the back.

Newton does nothing without intent, Leibniz told himself, and a sly thought lifted his mood. *Why not return Fatio's favour?*

Instead of writing a review of *Opticks*, he decided to review the calculus papers only – anonymously, of course – pointing out the similarity to his own previously published works on the same subject. Given the chronology of publication, who could say from where Newton had taken his inspiration?

Yes, thought Leibniz, *a parrying shot to finish the matter. Newton will understand the message and leave well alone.*

Dawn was leaking through the shutters when Leibniz put down his quill. His hand shook and his eyes smarted, but he felt buoyant.

He blotted the final page of his review and folded it into letter form, sealing the package with wax and addressing it. His back was stiff from sitting in the same position all night and he stretched his hands above his head. As he did so he remembered that he was supposed to present new pages of the wretched history before dinner that night. And, of course, there was Lady Caroline to see at nine o'clock.

He must snatch a few hours' sleep. He fumbled to close the book but it slipped from his fingers, falling open at random. As he lifted it, a phrase leapt out at him. He blinked at it, convinced his eyes were playing tricks.

The sensorium of God.

Something prickled inside Leibniz. *What did Newton mean by that?*

Unable to help himself, he hunched over the table and began to read the passage, then the following sentences and pages, and then the preceding ones. Time melted away. As he read, he lifted a hand to cover his mouth.

The sensorium of God.

Isaac Newton was peddling ungloved blasphemy to the English people as some kind of rational new philosophy. This could not go unchallenged.

There was a knock on the door.

'Lady Caroline? Is that the time already?'

Her pretty face was a mask of concern. 'Are you quite well? You do not look at all yourself.'

'I am of no concern, but you are just the person I would like to see now. Please, tell me what you think of this.'

She hurried to the table and he showed her the book.

38
London

The Royal Society's new home was a tall cliff-face of stone, hidden at the rear of a narrow crevice just off Fleet Street. Both sunlight and casual passers-by found it difficult to drop into the courtyard by accident, ensuring that the building, if gloomy, was mostly protected from the intrusion of the surrounding city.

Sometimes curious children would dive in to play there, but they were soon evicted if the President caught sight of them. Bellowed back into the streets, they carried tales of the ogre that lived in Crane Court, which inevitably brought more of their companions to explore the passageway.

There were no children today. Instead it was Halley's turn to suffer the ogre's rage.

'How can this be? How can this be?' shouted Newton.

Halley was sitting behind a desk full of Greenwich observations. Although he knew he was just the lens through which the President was focusing his anger, his heart-rate accelerated. He thought it best to remind Newton of the real villain. 'It seems that Mr Flamsteed is somewhat reluctant to stay up much past midnight. That's the only reason I can think of for the paucity of observations of the Moon's third and fourth quarters, and I can find no observations at all for one year.'

'An astronomer who cannot observe all night? Idiotic! How am I supposed to complete my lunar theory now?'

'There's only one way: a complete set of lunar observations over a Saros. My observatory in Oxford is almost ready.'

'A Saros?'

'Forgive me, it's my word for the nineteen-year cycle of the lunar orbits that Kepler noticed were linked to the tides. It's what I thought Mr Flamsteed's observations would provide us with.'

'Nineteen years!'

'But there is another way to serve your theory of gravity,' said Halley quickly. 'I have devised a crucial experiment. The comets of 1682, 1607 and 1531 all appeared from the same part of the sky, and they all followed similar paths. I'm computing their orbits to see if they are actually the same object, looping round the Sun time and time again. If they are, I will be able to predict its return. A fine experiment, do you not think?'

Newton showed no recognition of the question; something was at work deep inside him. 'Flamsteed has proved himself utterly incompetent, if not criminally negligent. Get him here. He'll answer for this. By my life, he will answer.'

A week later, Halley was pacing the flagstoned yard and wishing for the umpteenth time that today was not happening. He told himself that Flamsteed had brought it all upon his own head. Even so, he still wanted to warn the man of what he was walking into.

The noise of a carriage drew his attention, but his courage failed. What would Newton think if he happened to look out of the window and saw him talking to the enemy?

With a curse, he hid in the shadows.

It had been three weeks since Halley had looted the observatory, and the old man looked as if he had aged a year for every day since. He inched along, resting heavily on a walking stick, oblivious to his audience. If not for his portliness, Halley could have mistaken him for Hooke's crooked figure.

Wincing with effort, Flamsteed climbed the front steps. He fumbled with his walking-stick and it clattered to the flagstones. Paralysis finally broken, Halley rushed over and scooped it up.

'Let me help.' He took Flamsteed's arm, smelling the unmistakeable whiff of age and its thin masking of lavender water.

'Thank you, sir.' The voice wheezed with the effort. 'Are you a Fellow here?'

'John, it's me. Edmond Halley,' he said.

His quarry tried to snatch his arm away. Halley squeezed it more firmly.

'Don't. You'll crack your skull on the stone.'

'Summoned in the name of His Royal Highness Prince George, but I know who's really behind this fiasco. Two hours of torture to get here, forcing me to piss on the roadside because I can no longer wait out a single carriage ride.'

Halley flinched at the vulgarity.

'Shocking isn't it – to hear such words?'

'Don't lower yourself on my account, John. I know my manners have offended you in the past, and I want to apologise.'

'It's not your manners, it's your theft of my observations.'

'John, sailors are dying for want of your catalogue. You heard what happened last week. Two thousand lives lost on the Scilly Isles.'

'Poor navigation is my fault now? You're liars and thieves. All of you.'

'Let me take you for coffee afterwards. We have much to mend between us.'

'You and I have nothing to discuss as long as you align yourself with Newton.' Freeing his arm, Flamsteed struggled to open the hefty front door. 'You're nothing if you side with him in this matter.'

The door closed heavily, separating the men.

Newton contemplated Fatio's spidery handwriting as Flamsteed edged into the committee room. There was no return address and no signature on the letter. It had been hand-delivered to Newton's home and contained only a date and a time. He already knew the place. He kept the message concealed from the other committee members who were seated on either side of him.

As Flamsteed took his place, Newton folded the letter and tucked it into his jacket pocket. The Astronomer Royal was moving with such calculated lethargy it bordered on contempt. Newton did not wait for him to make himself comfortable. He picked up the committee's report.

As he spoke, Halley slunk into the room, his face downcast. He slid into a seat to Newton's left, near the panelled door.

'Reverend John Flamsteed, on the command of Prince George, this committee has reviewed your observations, taken during these past years at the public's expense. We have found you negligent and lazy in the collection of the lunar data, and tardy in the publication of this public property. Therefore, this committee has decided that your star chart is to be published without further delay.'

'You cannot force me to publish anything I am not ready to,' said Flamsteed. 'The observations are mine. I bought the telescopes when King Charles overlooked the fact that an observatory needs equipment. That makes them my property.'

Newton leaned forwards. 'You have been kept in employment these past thirty years out of the public purse. Your data is the property of the Crown, and we shall publish it as such.' He eased back, gesturing to his left. 'We have asked Mr Halley to make the final preparations and oversee the publication. Your further involvement is not required.'

'Halley?' gasped Flamsteed.

'Yes. This country's most distinguished astronomer.' Newton glanced to his left, where Halley was scrutinising the table-top, one shoulder turned away.

'I will not work with *him*.' Flamsteed jabbed a crooked finger.

'We're not asking you to. We have your data,' said Newton.

'He's a filthy adulterer and an atheist to boot.'

Halley cringed. There were indignant tuts from the others. Newton raised his voice. 'He is an esteemed member of this society; you will treat him with every respect. He will publish your catalogue, since you have proved incapable of doing so.'

'I will not take this from you, another atheist,' stormed Flamsteed.

'Atheist?' The word was a knife-thrust. Newton shot to his feet, the heavy presidential chair rocking backwards.

'We've all heard about your refusal to take Holy Orders,' continued Flamsteed.

'Reverend, you go too far,' said an outraged voice to Newton's right.

'Do I?' Flamsteed's eyes blazed. 'Ask him about his secret meetings with Locke. What did they talk about? We've all heard of the papers found in Locke's possession that spoke against the Holy Trinity. What's that if not atheism?'

Newton felt the papers he was holding crumpling in his grip. The others were looking at him. He cleared his throat. 'I will not be held accountable for Mr Locke's views. Your unfounded accusations serve only to divert us from the issue of your negligence with royal data. I warn you now, in front of witnesses, if you do anything to hinder the publication, you will be judged to have disobeyed a direct commandment from the Queen and Prince George.'

Flamsteed's face filled with a look of contempt. 'Your day of judgement is coming, and yours,' he said, flicking a look at Halley, 'but not yet from God. There are more earth-bound authorities that have their eyes on you and your irreligious thinking.'

Newton was still pretending not to have been hooked by Flamsteed's barbs when the old man was dismissed and shown from the room. The other Fellows made their own sheepish exits.

How had Flamsteed known about his religious views? Could he have read it in his eyes that day at Greenwich when he had questioned him about the biblical passage? Had he heard of it from gossip? Had he seen the anonymous pamphlet Newton had written for Locke and guessed the author?

'Isaac?'

The voice made him jump.

Halley was still in the room, his face drawn.

'Isaac, he may be right about the authorities taking an interest in our work. I've been visited, had questions asked about you and the theory of gravity. Some are saying it's irreligious thinking. *Are* you attacking religion with your theory of gravity?'

'What is it to you, Mr Halley? We all know that your beliefs are sadly lacking.' He charged for the door. He knew he had to get home and destroy anything that could implicate him.

Newton tore into the house and half-fell into his study. Dashing from pile to pile, he thumbed through ledgers, ripping out first one page, then another and another, and flinging them into the empty grate.

There was so much of his writing about this, how could he destroy it all? His entire work was shot through with the belief.

But I am still right on this one thing, rang his inner voice; *Christ was not fully divine.*

Where was the family Bible he had read, underlining the relevant passages to support his beliefs? Could he bring himself to burn the very book that had given his mother such comfort?

Gaol. He had some inkling of what it must be like from his visits during his war with the counterfeiters: the cold stone walls, the drip of moisture, the soiled straw, the rats, the cries of other inmates.

Even if he escaped gaol, he had heard of the lynching of heretics.

His desk was full of his attempted lunar calculations. Bellowing, he lunged at the sheaf of higgledy-piggledy papers, screwing them into big handfuls and dashing them in the direction of the grate.

But the room was still full of papers. He couldn't burn it all. It wasn't just his theology; there was his alchemy, too. That, in its own way, was equally incriminating. This room alone was filled with enough evidence to send him to gaol, maybe even hang him, several times over.

It came to him in a terrible realisation that there was only one way to stop this. If it was his theory of gravity that was causing people to ask questions, then he must publicly destroy it. And he knew how to do it. He had known it since the madness had left him and he had taken the job at the Mint. He had recognised it the very first time he had looked into the night sky with a clear mind and had seen the stars looking back at him.

A star had to be a burning Sun placed far off in space. So, each one must be producing its own pull of gravity. And that meant that each was pulling on the others, so they should be collapsing together – the whole Universe should be collapsing! It was pure, inescapable logic. Yet the stars were there, shining night after night, year after year, eon after eon, fixed in position, defying the very law of gravity he thought applied across the Universe.

Try as hard as he might, he could find no solution to this conundrum. There was no mathematical trick that he could pull. He was beaten, and not for the want of data this time. His theory of gravity was fatally flawed. All he needed was the courage to say so. That must silence his critics.

He looked around the room. A fresh panic engulfed him. Even if he spoiled his philosophy, the authorities could still find all the evidence they needed of his Arianism. Then no one would care about gravity anyway.

Think!

To his knowledge only two people knew for certain of his Arianism. His breathing began to steady. Yes, that was the most serious threat to him at the moment. The first was Locke, now dead and gone, and the second was Fatio . . .

Fatio!

What was the time? Newton was late.

As uncomfortable as it felt, Newton made himself look into the other man's eyes. 'Say that again.'

They spoke quietly, cocooned in the grubby clutter of Fatio's apartment.

'There is someone I would like you to meet,' said Fatio.

'Who?'

'A Frenchman working in London, Elie Marion. He's gathering true believers. His vision is yours: that someone is among us now who will bring us to the Kingdom of God. We must cleanse our souls in preparation. The Judgement is close at hand.'

'And you think this prophet is . . .' A bead of sweat trickled into Newton's mouth. He tasted salt and thought of Lot's wife.

Fatio's eyes were wide. 'You know who the prophet is.'

Newton felt queasy.

'Elie will know it too, as soon as he meets you. I've already told him about you. Will you meet him, Isaac?'

'How?' Newton had to buy thinking time.

'We gather in secret. Marion, me, the others.'

'How many others?'

'Every meeting there are more. Over thirty at the last.'

At that moment, Newton saw the solution as clearly as the dirty plate on Fatio's bedside table. He experienced a sense of awful inevitability.

'Promise me you will think about it,' Fatio urged.

Newton waited just long enough to make it seem as if he were weighing the decision. 'Send me the date and time of the next meeting. I will be there.'

39
Oxford

Although the banquet had yet to begin, the rafters were already echoing with wine-fuelled laughter and conversation. No sooner had the academics hoisted their glasses in a toast to the new King, George I, than the quaffing began in earnest. The dining-hall grew warm from the candles and the press of bodies.

Halley meandered through the gathering, stopping here and there to acknowledge a brief greeting or pleasantry.

How's the observatory coming along?

How are the comet calculations?

Have you seen the graffiti on the Arundel Marbles – shocking, isn't it?

I should say. I'd like to take the same knife that the villain used and use it on him – followed by a thunderous laugh.

'I simply cannot – will not – bring myself to acknowledge him,' said a diner. 'A Hanoverian duke is not a King of England. What say you, Edmond?'

'With the greatest respect, Mr Hearne, I know so little of these matters that I content myself to serve the monarch in possession at the time.'

A stung look greeted his comment. Halley thought it best to excuse himself and move on.

A few paces later, a nasal rasp made him turn.

'Couldn't have said it better myself. "The monarch in possession" – skilfully put.' Winslow smiled coldly. 'Oh, don't look so surprised. We old boys have to stick together, right?' He displayed his gowned arms.

'You studied here?'

'Awarded my degree for effort rather than by royal commandment. That embarrasses you, doesn't it? Don't worry, your secrets are safe with me. We're friends, remember, and friends don't keep secrets from one another.'

'Christ's wounds! There's nothing to tell.'

'That's not what the new Princess of Wales thinks. It seems that Princess Caroline studied under Leibniz and has some concerns about, shall we call it, Newtonianism.'

'Leibniz is nothing but a troublemaker.'

'Maybe so, but he's now the one with influence at court. I thought you and I both understood that survival is the only game worth playing?'

Halley looked away to the high leaded windows and the darkness beyond.

'Very well,' hissed Winslow, 'but don't ever say I didn't try to help you. You've never been to one of these gatherings, have you? Watch yourself: the entertainment can get a little personal, especially towards newcomers.'

Halley spent the next hour rolling tasteless food around his mouth; his mind did the same with Winslow's words.

After another toast to the King, the Fool took to the centre of the room. He was a young man of admirable height and musculature, with a flashing smile that drew in his audience. He was clad in the multicoloured patchwork of a mediaeval jester and carried an inflated pig's bladder that he used to dislodge a few wigs as he made a circuit of the room. There was a palpable expectation in the air.

He paused as he drew close to Halley. 'Gentlemen of Oxford, my task here tonight is to formally welcome our new Savilian Professor of Astronomy, Edmond Halley, into the bosom of the University. And, as you shall soon hear, if there is one thing that Mr Halley knows his way around, it's bosoms.'

Halley squirmed.

The Fool nodded appreciatively at the guffaws from the crowd. 'Doubtless you already know of those great services he has rendered to England: maps and voyages, star charts and so on, so I want to tell you of his greater labours. Yes, gentlemen, I know what you are thinking: what can possibly be greater than his work for the good of England? Let me tell you, it is his greater work to the population of the human race. Yes, this man has tirelessly sought to increase the number of humans in the world, selflessly giving himself to women over the years . . .'

There was another approving roar from the diners. Halley forced himself to lift his glass in salute, but his cheeks were burning.

'First, there was the lady of advancing years on his trip to Saint Helena, who suddenly found herself with child. Mr Halley modestly accredited it to the beneficial tropical air, but I think we know better, gentlemen. It was not the air but the vigour of youth – Mr Halley's youth, if I may call it that – that gave her a swollen belly . . .'

The Fool flashed a smile at Halley.

Halley remonstrated: 'That was her husband's doing, not mine.'

'Cheer up, Edmond,' said his neighbour, 'it's just a bit of fun.'

'There was also Lady Hevelius of Danzig.' The Fool strutted with open arms, then hugged himself tightly. 'Married to a much older gentleman. A gentleman, so rumour has it, whose great years kept his wife in permanent' – he opened his arms again and let his wrists flop lifelessly – 'frustration. In this situation, Mr Halley here did what any true gentleman would. He bedded her – and here's the best of it: he bedded her in her marital home. No clandestine tryst or rendezvous for him. He went to her very house and bedded her, while her cuckolded husband snored softly just a few doors away. Have you ever heard of greater devotion to the furtherance of the population?'

The sounds of hilarity bounced from the wooden panelling, but this time Halley detected an undertone of disapproval. Across the room one face stood out from the rest: Winslow's. There was no amusement on his face, just a steely look of satisfaction. The spymaster raised his glass.

The revels broke up around midnight, after the interminable music and poetry. Halley sat through the whole thing but absorbed nothing, his mood ruined. He refused every attempt to refill his wine glass and stared morosely at the residue of ruby fluid. When the boisterous etiquette allowed, he excused himself from the table.

The cool night air hit him as he entered the cloisters; he was glad to be away from the clinging fug of tobacco smoke in the hall. He breathed deeply.

'Edmond! A word.'

The torches were burning low but still offered enough light to reveal Winslow.

'This was your doing, wasn't it? You did all this just to humiliate me in public,' accused Halley.

'Tell me what I need to know about Newton and the humiliation will cease. I cannot convict him on the say-so of a German philosopher alone. I need corroboration.'

'There's nothing to tell.'

'You see, I've been thinking about you. What does an atheist have to fear? Not divine justice, that's for sure. No, it has to be something he'd miss while he still lives. So, I've reasoned that the thing someone like you would fear most is loss. Am I right? It could be loss of dignity, loss of reputation, loss of family.'

'I've warned you before, leave my family out of these affairs,' growled Halley.

'It doesn't need to be loss of life,' continued Winslow. 'This new lot don't go in for that kind of thing as much as the Stuarts did. But, of course, you know all about that, don't you?'

Father . . . The night darkened around Halley.

'But I digress. The loss could just be of trust, or respect, or perhaps love. Something that will last for the rest of a lifetime and hurt every single day. Say the loss of a wife's trust and love for her husband.'

The talking face filled Halley's vision. Then it was jerking upwards and backwards, a bubble of blood exploding from the lips as Halley's fist struck home.

'Leave my family alone,' he spluttered.

Winslow looked up from the ground, wide-eyed. 'Do you know, that's exactly what your father said just before . . . Well, you know . . .'

Halley launched himself at Winslow, fists clenched and ready to land more blows. He halted at the sound of footsteps and laughter echoing in the cloisters. Two men weaved round the corner, gowns askew.

'My, my, what's happening here?' asked one of them.

Winslow wiped the blood from the corner of his mouth and struggled to his feet. 'Nothing, gentlemen. I just lost my footing.' He straightened his clothing. 'And so has Mr Halley.'

40
London

Newton found what he needed in a squalid tavern on the outskirts of Seven Dials. Inside the pitted walls women went from table to table, pressing their breasts into the faces of the drunken patrons, snagging one after another to take upstairs.

Newton looked at the young man in front of him. One of his teeth was chipped, and the rest were stained and twisted. Occasionally the mouth shuddered into a cough, and Newton averted his gaze as the man spat on the tavern's earthen floor, peppering it with globules of sooty spittle. After each convulsion he dragged the back of his hand across his crooked nose and leaned back in his chair. His fingers were so dirty with coal dust they left black imprints on his tankard. 'I think we need to talk somewhere a little more private,' he croaked, 'but it sounds like thirsty work to me.'

Newton purchased a large jug of ale from the bar and followed the man to the rear of the tavern, where he pulled aside a ragged curtain and shooed away the occupants of the hidden booth. Newton filled his companion's tankard, then poured a splash of ale into his own so that he could pretend to drink.

'I'm told that you know what to do with heretics.'

The man displayed his teeth again. 'I keep the capital warm and cosy, and I don't mean just with coal.'

Newton grimaced at the juvenile patter. 'Have you heard of the Camisards?'

'Catholics?'

Newton shook his head. 'Prophets.'

'Prophets?' The man took a hefty swig of the reeking beer. 'You see, if they were Catholics, we could get a reward. You get the most for Jesuits, but hidden priests are pretty lucrative, too. Others though, well, no one's that interested in them. We just rough them up for sport, if we can be bothered.'

Newton hid his impatience. 'They're French prophets, plotting against the Crown.'

'Plotting?'

'They're spreading the belief that the monarchy is soon to fall and the Judgement will be upon us.'

'The Judgement.' The man whistled. 'I'm not interested in any biblical nonsense, but plotters . . . that's a different matter, especially with the changes to the Crown and all. Maybe a big reward.'

'They're certainly a danger to the new King,' coaxed Newton.

The man swilled beer around his mouth before swallowing with an extravagant smack of satisfaction. 'Right, when and where are they meeting?'

Newton stood up. 'Come, I'll show you.'

The location had been written on Fatio's last, anonymous note.

'Hold on, will you? I've to gather the rest of the lads first.'

An hour later, Newton pressed himself into the shadows. He tried to keep his shoes clear of the gutter and ignore the overwhelming stench of filth that pervaded this part of the city.

He could see the coalman's strapping team charging into the Red Lion. Shadows played across the interior of the windows. Newton watched for any hint of what was going on, half wishing he were in there to see the events unfold.

Through the door, hidden at the far end of the bar, Fatio's note had read.

Newton tried to imagine the surprise on the landlord's face as the squad barged past and into the secret place.

A sharp yell split the air and then there was uproar. Individuals dressed in dirty workers' clothes came streaming into the streets, shouting and grabbing at each other. Drinks were spilled and people cursed. There were a few skirmishes, but mostly they were running, plunging down the dark side streets, stumbling and falling in their panic to escape.

The pandemonium continued, with the tavern disgorging more people than it looked capable of holding. It was impossible to tell whether these were innocent drinkers – *although they could hardly be classed as innocent*, thought Newton – or the Camisards.

He took a step forwards when he saw the gang leader reappear. In his great arms Fatio was struggling like a slippery fish, cursing in French. He was followed by others from the gang, each with a captive. One, apparently unconscious, was being carried over a man's shoulder. As they drew closer Newton saw that all the victims were bloodied, with bruised eyes or swollen lips. He felt nothing.

The coalman noticed him watching and gave a curt nod. Fatio's damaged face caught the lamplight and Newton stepped forward involuntarily, into the captive's view. Fatio ceased his struggle, a silent question resting on his lips.

Newton watched.

The look of incomprehension dissolved and Fatio resumed kicking and screaming, not to escape but to claw his way to Newton.

'Traitor!' he screamed again and again, struggling so violently that the leader almost lost his grip. Newton stumbled backwards, suddenly fearful, before Fatio was restrained. The coalman barked to one of his accomplices, who stepped briskly forward. The crack of knuckles against Fatio's jaw filled the street.

The mathematician was silenced.

As the gang marched off to hand in their catch, a breathless Newton turned for home. The agitation in him softened with every step and by the time he reached the more familiar streets of the city he felt as though gravity was pulling on him less strongly than before.

Safe, at last.

Winslow marched down the stone corridor, holding a lighted torch. This deep in the Tower, the corridors were seldom lit. He ignored the black scurry of rats desperate to return to the cover of darkness and reached with his free hand to his aching jaw. It had been almost a week now since Halley had landed the blow, but the wretched thing refused to stop hurting. It ached most in the mornings, and made his nights restless.

He had wrought his vengeance, of course, as the astronomer would discover when next he visited home, but it brought little comfort. He hawked the last of the phlegm from the back of his throat and spat it on the wall. He was getting too old for this kind of life.

A slumbering guard came into view. Winslow paused a moment before kicking the man's foot.

'I'm told you've got new plotters. I want the one who claims to know Newton.'

'Yes, sir,' said the embarrassed guard, hurrying to unlock a cell door.

Inside the cold space Winslow could hear the whisper of the Thames, disconcertingly close, and the steady drip of condensation. The prisoner shot to his feet so quickly that Winslow stepped back, fearing attack. He brandished the torch and the man stopped moving. In the orange light Winslow could make out lank hair tangled with straw and a face covered with bruises and dried blood. The jaw was grossly deformed.

'Who are you?' he asked.

'Nicolas Fatio de Duillier.'

'You claim to know Newton.'

'It's him you want, not me,' Fatio slurred.

'Why?'

Fatio drew his arms around himself.

'Come now,' said Winslow, 'this is no time to be coy, even if it does hurt to speak, especially after what you are charged with: plotting against the royal family.'

'But he's the one plotting – against the Church.'

Winslow nodded appreciatively, then called over his shoulder. 'Guard, fetch this man some washing water, and a drink.'

A bucket of water and a mug of thin beer appeared.

Fatio downed the beer, then washed his face, wincing at every touch of the rag. 'Thank you.'

'Do you know, I think that jaw may be broken. The beer will numb it, but I could fetch a surgeon to look at it. Of course, we'd need something to talk about to pass the time before he arrives.'

Fatio nodded mutely.

'Good. I know it will hurt, but, be a good chap and start from the beginning. Tell me the whole story. Where did you and Newton first meet?'

Fatio began his tale, pausing often to swallow and wince.

Winslow called for a stool to sit on. This was going to take all night.

Hours later, the guard came back into the cell with two thick-set colleagues. 'It's time, sir. I'm sorry.'

'It cannot possibly be morning already,' said Winslow, more for show than anything else. He already had enough ammunition to take to Lady Caroline.

'The light doesn't tend to reach down here, sir, but it is time.'

'Where's the surgeon?' asked Fatio, looking past the men into the corridor.

Winslow gave a small shrug. 'Too late for that, I'm afraid.'

Fatio's face filled with horror. He pressed himself into the corner of the cell and sank to his knees.

The guards drew nearer.

'Wait!' screamed Fatio, waving his arms above his head. 'Wait!' His eyes were crazed and his temples throbbed. The stench of urine filled the air.

Hauling him to his feet, then pinning his arms behind his back, the guards forced the screaming man from the cell.

'Don't I even get a holy man?' pleaded Fatio. 'I must be allowed to pray. I cannot meet my Maker like this.'

Winslow laughed. 'You don't need a holy man where you're going.'

41

Halley's fatigue from his early start vanished the moment the pamphlet hit the wall. As it smacked against the pale wash, the glue gave way and pages rained down on to the floor. A moment before, it had been in Newton's hands. Next to him was a young man with auburn hair and freckles, looking stunned. He had appeared thoroughly pleased with himself when bounding down the staircase from the Society's library, waving the document in his hand before handing it to Newton. Halley wondered whether this was the first time John Keill had witnessed the President's temper.

Around them sweaty workmen carried in crate after crate and stacked them in the Royal Society's airy lobby. Each wooden case contained fifty copies of the Greenwich star catalogue, *Historia Coelestis Britannica*: four hundred printed in total, seventy already reserved for purchase. The delivery was the reason Halley had dashed from Oxford.

Keill rushed to gather up the fallen pages. 'But it's an anonymous review,' he said in a squeak of a voice.

'It's the work of Leibniz! Trust me, he's one of the journal's editors. Even in the unlikely event that it was written by somebody else, he sanctioned it. The traitor!'

Halley knew better than to ask, but it did not stop him. 'What has he said?'

Newton waved his hand at Keill, who passed over the loose pages. Halley read the Latin with disbelief. The review implied that Leibniz was the first inventor of the calculus, and that Newton had stolen the method for his own fluxions.

Halley looked up at Newton's thunderous face. 'This is ludicrous – erroneous in every regard. Leibniz knows the truth, because he saw your papers on power series when he visited the Society in the 1670s.'

Newton's face darkened further. 'Leibniz saw my papers? How do you know this?'

'He told me when I was in Hanover.'

Halley thought Newton was going to explode. The President screwed his eyes shut. 'So, Leibniz saw my papers, but chooses not to acknowledge seeing them in his own. Who is the plagiarist now? The Society must convene a committee to report the facts of this matter and publish the precise chronology of the inventions. And it must do so quickly. Second inventors count for nothing, as Mr Leibniz will very soon discover.'

Halley's thoughts filled with images of the isolated man in the library at Hanover. He looked back at the paper. *Gottfried Leibniz, what have you brought down upon yourself?*

Halley sensed an atmosphere the moment he arrived home. William was hovering in the hall and the house was unusually still.

'Mrs Halley is in the drawing-room, sir. She would like to see you in private.'

Halley paused long enough to realise William was not about to give him a clue as to the trouble before hastening to his wife.

Mary did not acknowledge him but maintained a steady gaze on the window to the street.

One of the children?

'What's wrong, my love?' Halley rushed to her side.

'My love? You have the nerve to call me *my love*.'

He knelt beside her, looking at her profile. There was a film of moisture in her eyes that caught the daylight. 'Is everything all right? Are the children all right?'

'I think it better if, from now on, you live permanently in Oxford.'

Halley shook his head. 'Please, Mary, what is the matter?'

She was dressed in dusky blue, a colour that Halley loved. He wanted so much for her to turn round and look at him. 'Who is Lady Hevelius?'

Halley went hot and cold. 'Who has been here, Mary?' he demanded.

His anger forced her to look at him. Her eyes were red-rimmed and ruinously bloodshot, but her flinty gaze drove fear deep down into him. Never before had he been so frightened, not even when the bow of his ship had crested storm waves and dipped so low that he had looked down into the black, swirling ocean.

'I begged you, the morning after that awful party: anything but other women. I implored you not to turn me into a laughing-stock.'

'Mary, I haven't. As God is my witness, I haven't.' He wanted to seize her hands, but he dared not.

'God? We both know that's an empty promise from you.'

'That's not fair. I've taken the family to church every Sunday since my return. I am trying to believe.'

'Now I think back,' she continued, her voice starting to crack, 'I can see what a fool I've been. The sudden visits, supposedly to your father; the coffee-shop meetings and the conversations; the voyages; the trips to Cambridge. I trusted you. You convinced me you were not like other men. Oh, how foolish I've been! You and your women – it seems all London knows of it, except me.'

The urge overwhelmed Halley and he grasped her hands. He drew them tightly together. She struggled to free herself, but he moulded them into a single fist and gripped them in his own, as if they were praying together. 'Mary, my love, I have had secrets from you but not about other women. I promise you, I have been in trouble but not with other women. My secrets were to protect you. I think you have been visited by someone. Someone short, with little eyes, who speaks in a nasal voice.'

A sidelong glance told him that he was correct.

'He's called Winslow, and he's out for vengeance because I will no longer help him.' He spoke in a rush, the force behind his words having built up over decades. He told her about his father and the Earl of Essex, and watched some of the anger drain from her face. 'I couldn't tell you any of this at the time. The knowledge was too dangerous.'

'I thought your father was murdered by thieves.'

Halley shook his head.

'But how can we be sure that this man won't try something similar to you?'

He took her hand and kissed it gently. 'He would have done so already. I'm no threat to the monarchy, just a thorn in his side. That's why he's planting this mischief. He's losing power and wants to drag others down too. He's obsessed, Mary – he thinks philosophy is a danger to religion and government.'

'Stop right there.' She pulled her hands away from his. 'I know exactly what you're doing. You're diverting me from your adultery. I'm an object of ridicule, Edmond.'

'What adultery?'

'With Lady Hevelius. It's a mortal sin, Edmond; it's in the Ten Commandments. How can you come to church with this in your past? How can you be so bold?'

'But nothing has passed between Lady Hevelius and me.'

'Half of London is apparently talking about it.'

Halley inched nearer. 'Let me tell you the truth about Winslow, my service to the Crown, my trip to Danzig . . . everything.'

Shadows had replaced sunlight by the time Halley's story drew to a close. As the light had dropped so had his voice, and the two of them had been drawn inexorably closer. They were leaning towards each other now as he all but whispered his final confession.

'It is true that Lady Hevelius came to my bedchamber. She asked me about London fashions and wanted to have a dress made out of the latest materials. I agreed to buy the fabric for her. I don't know how the story became so exaggerated. I hope that it did not begin with a mistaken belief on Jan Hevelius's part. I promise you faithfully that I have not made a fool of you, or myself, with other women. You're the only woman I want in my life. You're the reason I gave up sailing. I've been away from you for too long. I've missed you too much, and . . . and I want to make the most of you while we both still live.'

'I want to believe you, Edmond, truly I do. But I've heard that Mr Flamsteed believes this story. Why would he spread such hateful rumours?'

A fatigue settled on Halley. 'I've hurt him grievously by publishing his star catalogue. Perhaps I've taken on too many of Isaac's fights as my own, I don't know. I regret what I have done to John; he was such an early inspiration to me. Now he's old and sick. I fear he will go to his grave hating me.'

'Then do something about it. The man I married solved problems, he didn't wallow in them. Let me see that he still lives.'

Halley looked up, jolted by the challenge in her voice. 'I will, Mary,' he said. 'I swear I will.'

The following morning, Halley lined up his family in the sunshine, inspecting them as he used to inspect sailors on deck. They were standing along the red-brick perimeter wall of the observatory at Greenwich, all wearing their Sunday best. Edmond junior was standing to attention.

'No need to go that far,' said Halley mildly, and his son relaxed. 'But there are to be no jokes today,' he said, catching Katherine's eye.

Her older sister nodded in agreement.

Halley turned to the person on the end of the line, dressed in black jacket, breeches and a clerical collar. 'Thank you again for coming with us today, Reverend Hargreaves, especially at such short notice.'

'It's the least I could do for you and your family.'

Halley had caught the new vicar gazing at Katherine on the carriage ride. His interest had not gone unnoticed by her, either.

'Tell me, Reverend Hargreaves,' she had said, 'do you notice the way the young men and women of the parish catch each other's eyes in church?'

'Katherine!' Mary had interjected, sparking a giggle from her daughter.

Hargreaves had blushed furiously.

Now, standing next to the family, he said, 'Though, I confess, I didn't realise the observatory was so far outside London.'

'It has to be far enough away so that the city fog doesn't obscure the view. The ordinary clouds are nuisance enough.' Halley led them through the entrance gate and into the small courtyard. As he paused before the steps, Mary brushed his shoulders clean of wig-powder and gave him an encouraging nod. He stepped to the door, knocked and stepped backwards into the midst of the gathering, adjusting his clothing yet again.

It took an age for the door to creak open. As it did Halley fought a surge of unease. Bringing his whole family here – and a clergyman – unannounced was so obvious. He quelled the feeling; of course it was obvious, it had to be recognisable as a gesture of apology and atonement.

Flamsteed gawped at them round the door.

Halley looked back.

'Reverend Flamsteed, a pleasure to meet you,' said Margaret in her serious tone. 'Father often speaks of you.'

'John,' Halley said, but stalled after the first word.

Flamsteed cautiously opened the door. 'You'd better come in.'

Halley ushered his family and Hargreaves into the observatory building. Mary smiled hopefully as she passed.

There were not enough chairs, so Halley's children and the vicar pressed themselves against the sitting-room walls, like medical students at their first dissection. Flamsteed and Halley sat at the small central table. Mary perched on a stool.

'I keep a simple home,' said Flamsteed, unabashed.

'Is Margaret here?' asked Halley. Turning to his elder daughter, he said, 'You share his wife's name.'

'Derived from the Greek word *margarites*, meaning pearl,' she said.

Flamsteed ignored the exchange. 'She's staying in the parish this week. I'll be joining her at the weekend.'

Halley remembered the time, long ago, when he had come here convinced that Flamsteed's rectorship would make vacant the post of King's Astronomer. The rebuff still embarrassed him. Realising he had slipped into reverie, he rubbed his hands together, looking around for inspiration. Hargreaves was looking at him expectantly.

'I've been telling the Reverend Hargreaves about a most curious calico shirt that I saw on the docks,' said Halley. 'It was brought from India, I discovered. It was woven in such a way as to be entirely without a seam. Had I not seen it myself I would have never believed it.'

Flamsteed stared.

'It explains the scriptural revelation about our Saviour's coat being seamless.' Halley could hear himself floundering.

'Oh yes, we were discussing it on the coach ride over here,' chipped in Hargreaves.

'You came by coach?' asked Flamsteed.

'The six of us were pressed so tightly together it prevented us being thrown around by the potholes.'

Flamsteed did not react. Halley's smile died.

'Let me show you the errors I have noted in your star catalogue.' Flamsteed huffed from the table and shuffled into the other room.

When he returned he dropped the book on the table, forcing Halley to snatch his fingers away before the tome hit them. 'In truth, there are only a handful of pages that are free of error.'

Halley dropped his gaze. He knew that he had not paid as much attention to the work as he should have, but that statement must be an exaggeration.

'I'm told that you were paid for your editing, Mr Halley. Yet I receive nothing for my preparations . . .'

Except your salary, thought the younger astronomer.

'. . . nor has my Margaret, who has learned enough mathematics to be an invaluable assistant.' The disgust in Flamsteed's voice was obvious.

Before Halley could say something he would regret, he reached for Mary's hand. 'I understand the value of a good wife.'

The bloated face remained impassive.

Halley lightly touched his fist to the table. 'Heavens, John! Believe me when I say that I would gladly burn every copy, if you would just publish your version.'

Flamsteed's face twitched. 'Burn every copy? A sacrifice to heavenly truth?'

'If you like,' said Halley.

'If I were to burn this volume and publish my own, I wonder if Sir Isaac – or you – would recognise the kindness I have done?'

'I would be the first to applaud you for your efforts.'

The family and Hargreaves left soon after. They gathered outside in a similar fashion to their arrival, and Halley looked from one uncertain face to the other.

'How about repairing to Greenwich village to find some lunch?' he suggested.

His son agreed immediately, needing no further prompting to turn down the hill. Everyone followed. Hargreaves and the children were soon a dozen paces ahead, Katherine every now and again gently bumping shoulders with the clergyman.

Halley ignored her flirting and looked back over his shoulder at the red-brick observatory. 'He's as immoveable as the bend in the Thames.'

'At least you tried,' said Mary, slipping her arm through his. 'He may yet come round.'

'I once fancied that I might succeed him. Now . . .'

She drew him closer. 'Let's not talk about the future. It will unfold all too quickly and of its own accord. But we have this lovely afternoon together. Let's enjoy it.'

Her proximity reached into his core. 'Come on,' she said, urging him into a run to catch up with the others. It was an ungainly feeling to start with, but with each footfall he remembered the joy of running and wondered why he had ever stopped. Laughter welled up and broke free, drawing the attention of Margaret, whose mouth gaped in surprise at the sight of her old father careering towards her.

Back in the city, a huddle of prisoners stood on a raised platform. It was evident that they had been there for some time. Bent over, their heads and hands were poked through filthy wooden stocks; their garments were streaked with food and other stains. Up high, the sun beat down its own humiliation.

Newton stepped towards the criminals.

Fatio was in the front line, where they put the worst offenders so that they would take the brunt of the public's assault. His forehead was red and angry from the sun. His hands looked empty of blood and he stank of rotten food and bodily fluids. His eyes were slits. He swallowed before uttering the words: 'I wish they had killed me.'

Newton made a disappointed sound. He had not come to hear self-pity. Now he thought about it, what had he come here for? Something had compelled him to make the journey from the Mint that morning. Perhaps it was just to see the end of this particular story. Well, he had seen it now in all its ignominy. He made to leave.

'Wait,' croaked Fatio, bringing Newton to a halt. 'I loved you.'

'You loved me for what you thought I could offer you and your foolish ideas.'

'All I ever tried to do was marry your new philosophy with the old.'

'There is no marriage to be had. Everything is different now. The best thing you can do with your manuscript is to burn it.'

There was something in Fatio's gaze, something so unfathomable that Newton decided it was further evidence of the man's cracked mind. He turned again to leave.

This time Fatio made no attempt to stop him.

42
Hanover

Leibniz tried to stop the tears.

Newton has finished me.

The words echoed in his head, drowning out the litany coming from the petty official standing over him. The young man had not even been bothered to deliver his verdict in a private office, preferring to ambush him in one of the state rooms, within earshot of the staff who were busily packing more of the King's possessions for transportation to London.

The throne that I helped negotiate for him, thought Leibniz bitterly.

'You do understand what I've been saying, Mr Leibniz? The evidence against you is compelling.'

'I did not steal Newton's mathematics.'

'We have an official document, authored by independent advisers from the Royal Society.'

'Independent! They're his mouthpieces.' He raised a fist.

'The Royal Society says that you were shown Sir Isaac's private papers before you published your calculus,' intoned the official.

'But I could take nothing from his work that I didn't know already. It only proved to me that we were both upon the same subject. How can I make you understand? I'm cut adrift. If only I could accompany you to London, I could see Newton and reconcile this matter.'

'I have just told you that the King desires you to stay here and finish the family history.'

'But I could write about a new history of England.'

'He does not want you in London.'

'If I could just see His Majesty–'

'He is busy with more important matters. Compose yourself.'

Leibniz sensed a tear rolling down his cheek. All his life he had imagined joining the meetings at the Royal Society, watching the experiments, reading his papers and applauding others for their efforts, contributing. But now, on the brink of it, Newton had struck a fatal blow. He was to be excluded utterly and for ever. Not even his own prince would support him.

The official screwed up his nose as Leibniz tottered pathetically. 'You do see, don't you? Sir Isaac Newton is the most important philosopher in the world. You are nothing but a court historian who's been unable to finish a task. You've merely dabbled in mathematics, and now you've made the mistake of borrowing from the greatest living mathematician. It's impossible for you to join the King in England. You'd be an embarrassment.'

'Forgive me,' said Leibniz stiffly. 'I have work to do, for the King of England.'

All along his route to the library, he noticed the indentations in rugs where furniture had been removed and the bright squares of paint where pictures had been unhooked. He paused for breath by a window and looked out over the garden; the hedgerows and flowerbeds might just as well be prison walls. As he leaned on the alcove, an idea stirred. It was fully formed. Perfect.

He might not be able to challenge Newton in person, but Princess Caroline could. She was already in London, going about her duties as the new Princess of Wales. Had she not been equally appalled by Newton's tinkering with the meaning of God?

He recalled her outrage the morning he had shown her Newton's *Opticks*. All he had to do was write her a letter.

It could be in the hands of the courier tonight and in her own within a week. Reminded of the damage one of her subjects was perpetrating, she would feel duty-bound to investigate.

A grim satisfaction took hold. One letter, and he could set Newton's downfall in motion.

43
London

Halley was in the mood to celebrate as he squeezed into the Waggoner's. It was loud and sweaty. Unlike the coffee-shops, where the conversation was generally intellectual, here men talked for no other reason than to amuse each other or brag about their exploits.

Perfect, he thought, wanting to prolong the visceral sensation of triumph that had exploded inside him that afternoon when the numbers had fallen into place. Racing from the Society's library, he had paced across Crane Court a dozen times, tracing out the shape his mathematics had revealed.

It was a perfect ellipse, long and narrow, and it joined up every observation of the comets of 1531, 1607 and 1682. Inescapably, these three objects were but a single one, endlessly travelling around the Sun. At first it would fall inwards to whip round the Sun, then it would coast back out into the inky darkness way beyond Saturn before the Sun's gravity inexorably pulled it back in again, decades later.

It was out there now, somewhere beyond the reach of any telescope, guided by gravity. *But I can calculate its position and know where it is*. He relished the godlike thrill of the knowledge, of the certainty he could have of nature, all of it made possible by Newton's understanding of gravity.

He soon found sailors to drink with, mostly new arrivals off a ship bringing sugar cane from the West Indies. They were eager for three things: to drink, to find women and to set sail again. As refreshments kept arriving, so the men talked more about the women they wanted to find, but none of them seemed to be doing anything much about it. So they drank more.

'I can tell the future, you know,' Halley remarked to a craggy, one-eyed sailor.

'Can you now? I suppose you'll be asking me to cross your palm with silver next.'

'No,' said Halley, 'not like that. Calculated, scientific prediction. I've studied the paths of comets and found one that returns from the same part of the sky and follows the same path every seventy-six years or so.'

The sailor searched for words, eventually plumping for a bemused 'Well done.'

'It will next appear in 1758.'

'Can you also tell when the women'll be here?'

'You don't understand. If I'm right, I have proved for all time the value of Isaac Newton's work on gravity. I have provided the theory with a crucial experiment–'

'I think you need another tot, my friend,' laughed the sailor, revealing more gaps than teeth. His compatriots were swift to join in. Soon Halley was smothered in a friendly chorus of derision.

'I need my wife, gentlemen. I bid you good-night.' He headed for the door, weaving only slightly: nothing a quick nap in the carriage on the way home would not fix.

He had not walked ten paces down the street when powerful arms encircled his neck and gagged him from behind. A black hood was pulled over his face and his legs were kicked from under him.

Halley knew he was being taken to the Tower when the repetitive pounding of the Mint's machinery reached through the hood. Terrifying images of the dungeons crowded his mind.

When he stumbled, a powerful hand clasped his arm and then marched him down corridors, staircases and passages so narrow that his shoulders rubbed the walls. He heard the creak of a door opening and he was planted in a rickety, straight-backed chair. The hood was jerked from his head.

Shielding his eyes, he squinted at his new surroundings. He saw soft furnishings, polished chests and vases, candelabra, an old Yeoman Guard, and Winslow, sitting not six feet away.

'Do make yourself comfortable. There will be plenty of time for chatting later, but right now I have to slip away and invite another . . . "guest" to join us.' He stood, sweeping his gaze around the plush room. 'I thought you might like it here. It's one of the state rooms James II used when he was the Duke of York. In fact, it's the room where I first encountered your father.' He winked and was gone.

Newton had long since learned to filter out the crump of the Mint's machinery. It carried so easily through the wooden walls that he could easily pick out the coin-cutters from those stamping the patterns on the face and from those milling the edges. Against this percussion, he was signing off the latest batch of invoices for raw metal when a new sound drew his attention. Rhythmic footsteps made their way along the corridor's floorboards.

A stranger entered the room. He was a short, stringy man clad in loose black clothing who wore an expression so cold it took even Newton aback. He spoke in a voice that seemed to lodge in his sinuses.

'Sir Isaac Newton?'

'Who wants to know?'

Winslow gave him a withering look. 'Please be seated.'

'I prefer to stand.'

Winslow shrugged. 'As you wish. What's your association with Nicolas Fatio de Duillier?'

'Just who are you to force your way in here and ask such a question?'

Winslow's gaze held steady. 'I can assure you, I have every right.'

Two red-jacketed Yeomen entered the office.

'They're more for show than anything, I admit, but it would still be best if you answered the question,' said Winslow.

Curiously, at that moment Newton thought he could smell the boiled bacon he had eaten for supper last night.

'Now, where were we? Fatio de Duillier.'

'We were acquaintances a long time ago,' Newton forced himself to say.

Winslow issued a faint noise, its meaning impossible to discern. He nodded towards Newton's wig, resting on its stand. 'You might like to bring that.'

Newton did nothing.

'It's up to you. But you will follow me, in the name of the King.'

Newton stepped round the desk, lifting the wig to his head as he passed Winslow, slowly and deliberately. He could see clear across the top of Winslow's head as he followed him along the corridor away from the Mint. The Yeomen followed several steps behind. Neither would look at him, but the steady tread of their steps helped focus his mind.

He had prepared for the possibility of this day. He would reveal the disastrous flaw at the heart of gravity. That would be enough to satisfy them. He forced himself to walk at a slower pace, although every nerve-ending was sparking. He began to rehearse his speech.

There was an eerie whisper from a multitude of internal voices. Newton felt a tingling down his spine. It was a revelation, if not an epiphany. *Maybe the flaw was not a flaw after all.*

Winslow led them outside, and the Tower walls came into view around them. Newton blinked in the bright light. *Think fast.*

The Yeoman Guard was looking at Halley incredulously.

'You look as if you've seen a ghost,' said Halley.

'I feel as if I have. You must be Edmond's boy.'

Halley froze. 'You knew my father.'

'I'm Thomas Redman. I was little more than a boy at the time. Haunted me for years what they did to him, you know. What *he* did to him.' He threw a glance towards the door. 'Your father was an honest man, Edmond. He didn't deserve any of it.'

At that moment Halley heard some movement in the courtyard. He crossed to the window. Through the uneven glass, it looked as if crimson and black chemicals were sinking in blobs to the bottom of a flask.

More guards . . . Winslow . . . *and Newton!*

When Newton was led into the state apartments he was still thinking hard.

'Wait here,' said Winslow. He disappeared into an adjoining room and appeared moments later with Halley in tow. 'No time for a reunion, I'm afraid. Someone has asked to see you both.' He led the two men deeper into the building before bringing them to a halt before a pair of panelled doors and gesturing with his right arm. 'Do go in.'

Newton led the way. A hooded figure was sitting at the far end of the room, silhouetted against the window. There was a guard in each corner of the room.

'Sir Isaac Newton and Mr Edmond Halley, Your Royal Highness,' Winslow announced.

The figure reached up and pushed back the hood, revealing an abundance of flaxen curls crowning a high forehead. Her milky skin contrasted with the black of her cloak. Tawny eyes radiated confidence.

From the look on Halley's face Newton deduced two things in quick succession: that the astronomer knew her, and that he and the astronomer were in trouble.

She placed her hands in her lap as though posing for a portrait. At her side was a copy of *Opticks*, from which thin fabric strips dangled.

'I don't know you,' said Newton.

Winslow scowled. 'This is Her Royal Highness, Princess Caroline, the Princess of Wales.'

Newton mumbled an apology.

'Your Royal Highness,' said Halley, bowing from the waist, 'a pleasure to see you again.'

'Gentlemen, thank you for coming today.' Her voice betrayed no warmth or intention. 'I'm aware that many men hold you in the highest esteem, and that my predecessors also held you in regard. Nevertheless, I have questions; particularly for you, Sir Isaac.'

Winslow stepped forward. 'Ma'am, Mr Halley was the editor and publisher of the book that started all this, Sir Isaac's *Principia*.'

She gave the slightest acknowledgement, keeping her focus on Newton. 'Someone whom I respect greatly for his learning and wisdom has brought to my attention certain matters that are most troubling to me. I admit that I'm still finding England new and perhaps a little strange. The thinking and the customs are different, but one thing I would never be able to adjust to would be a country that does not hold the highest moral and religious principles. We are here today to establish whether your experimental philosophy is leading this country into a state of theological decay. Tell me, what does God mean to you?'

'God is our supreme Creator, the fountainhead of all knowledge and love.'

'And yet by your own printed words, you seek to diminish the power of God.' The stark challenge in her voice was evident.

'In no way whatsoever–'

'Forgive me, but how do you explain the passage in your latest book that talks of God's need for senses to perceive the universe?'

Newton made to speak, but she raised her voice. 'Do not think of contradicting me. You call the Universe the sensorium of God. You suggest that we are all figments of God's imagination. That we are somehow held as illusions in his sensorium.'

Newton steeled himself. 'You have been schooled by Leibniz, but may I remind you that the Royal Society has found him a plagiarist and a liar? He has yet to answer these most serious of charges.'

'I have known Mr Leibniz for much of my life and I have found him to be neither a plagiarist nor a liar. You will treat him with respect.'

Newton brought his hands together. 'I fear you take my inadequate words too literally over the sensorium of God. Look again and you will see that it says *as if it were* the sensorium of God, rather than a definitive assertion.'

'Are you calling Her Royal Highness a liar, Mr Newton?' Winslow growled.

'Never.'

'I can assure you that the copy I have seen has no such words,' she said.

Newton opened his liver-spotted hands. 'Then I admit that, by a careless pen-stroke, I omitted those vital words in my first manuscript. While the book was in the press, I realised my mistake and submitted the change to be corrected. We had the page cut from the book and a replacement stitched in. Examine the copy on my shelf and you will see. However, I fear that a few copies were dispatched before the correction was made. You may have seen one of those false volumes. Unlike our Lord, I am fallible.'

'The excuse makes no difference,' said Winslow. 'You were a friend of John Locke's.'

Newton's skin prickled into gooseflesh. 'I knew of him.'

'You spent time with him at High Laver a number of years ago.'

'What of it?'

'His papers reveal heretical thoughts about the divinity of Christ.'

'I am not Mr Locke.' The room seemed to be shrinking. Winslow was definitely closer.

'But you have had revolutionary thoughts about religion, have you not? Tell us of your association with Mr Fatio and the Camisard prophets.'

Newton found words impossible.

'Then let me remind you,' said Winslow. 'The Camisards are a group of French Huguenots and plotters. They've been gathering followers, claiming that the monarchy is about to fall and will be replaced by the divine Saviour. But, I ask you, who is the Saviour?'

'All you need to know is that I was the one who exposed the Camisards. Would I do so if I shared sympathies with them?'

'Quite so, sir,' said Caroline before turning to Winslow. 'That is quite enough of your conspiracies . . . Sir Isaac, you still have not explained why you think God needs senses to perceive His own creation. The mere suggestion sets the two apart from one another and denies God's omnipotence.'

'I never intended God's sensorium to imply that He needs eyes and ears – all of those are material objects, and God cannot be material like the rest of his creations. He must be as pure as the tiny part of us that thinks and reasons. Our sense of identity and our reasoned perception are not flesh and blood; they are God's gift to us. They raise us from the animals and provide the essential link between our Creator and us. Using His gift, we can recognise Him in everything around us. If not for God's design, the creatures of the Earth would be hideously misshapen – here a useless arm, there a redundant lump of flesh. One needs only to look at the phenomena of nature to see God's work. Even if there had been no revealed truth and therefore no Bible, my treatise shows that we would know that God exists. This is what I mean by calling this great Universe His sensorium: that we would know the presence of His supreme intelligence because our experimental investigations show the design of nature.'

Princess Caroline studied him. 'You believe the Universe is a mechanical place, do you not? A place governed by the laws of gravity and motion?'

'Yes, I do. Fashioned by God's own design,' Newton said with growing confidence. 'My works in this regard can be seen as new gospels–'

'Then you must have a poor opinion about His capability. Have you not written that your law of gravity means that the planets will nudge each other from their orbits, and that our Lord will be forced to rearrange them back into the perfect order?' She lifted the book and removed one of the fabric strips. '*He may fashion and refashion all parts of the Universe as he chooses*.' She lifted her eyes to look straight at Newton. 'It is a poor clockmaker, is it not, who is forced continually to repair his creation, and doubly so if he has to do it with miraculous interventions? For anything that goes against the natural laws must surely be a miracle, and miracles must surely be actions of God's grace, rather than necessity.'

Newton was again at a loss for words. How was it possible for a woman to do this to him?

'Well, Sir Isaac,' she said in her deceptively soft voice, 'is God not capable of the divine wisdom to make a perfect creation?'

Halley, too, was looking at him expectantly.

Newton mustered his courage. 'Who are we to say what is perfect? God does everything for the best – even Leibniz believes this – but what constitutes *best* is for God alone to know. We cannot presume to know His mind. The mechanical laws are just one aspect of this Universe; His divine will is another. I'm no further towards understanding the cause of gravity. I can describe it from the phenomena I see around me, and that elevates it from being occult, but its cause remains beyond any mechanical explanation. Gravity may even be the will of God Almighty. I study two sides of the same coin: natural laws on one, God's will on the other. That which I cannot understand must be God's will.'

Princess Caroline opened the book again. 'You write in another passage that the sensorium of God *is analogous – though to a far higher degree – to the way in which our souls, which are the very images of God implanted within us, are capable of moving the parts of our bodies at will*. This strikes me again that you are clearly asserting God is a material entity, with the various celestial objects as His limbs.'

'On the contrary, Your Royal Highness. That is not my intention at all.' Newton fingered the curls of his wig, then blurted, 'There is a flaw at the heart of my theory. It is heinous to my reputation, but I must tell you, because through it I have found God.'

'Found God?' growled Winslow.

'Proved God!' Newton clasped his hands in supplication. 'You will never again say that my work is irreligious if you will just hear me out.'

He looked from one to the other.

'Continue,' said Caroline.

There was twisted glee on Winslow's face. 'Yes, continue, Sir Isaac! At last, tell us of your great revelation, about how you think yourself the equal of our Saviour, Jesus Christ. Did you or did you not confess this to Mr Duillier?'

They were flies trapped in amber. Halley looked at Newton in horror; his arguments had sounded increasingly desperate, and now he had stopped talking altogether.

'Did you not confess this to Mr Duillier?' pressed the spymaster.

Say something, Isaac!

Halley stepped forward and addressed Winslow. 'You pilloried Fatio for his crimes. You can hardly claim him as a reliable witness.'

'Did you or did you not confess it?' insisted Winslow, ignoring Halley's outburst.

Halley persisted. 'If you were any good at collecting intelligence instead of pursuing your own petty prejudices, you would have discovered that Sir Isaac laboured under a fearsome distemper in 1693. It made him shout out many strange notions. We feared for his life and his sanity. He wrote letters to people – Samuel Pepys, for one – letters that made no sense, accusing these men of dreadful things. They at once alerted us to the fact that he needed help. If Monsieur Duillier also heard something out of turn–'

'Once I was rested, my shameful notions passed,' Newton cut in.

'I've heard different.' Winslow's sour breath touched Halley's face.

'What makes you always right, Mr Winslow? For decades now I have served the monarchy of England faithfully through you. I have done everything that was asked of me without question. But now I want to know, what is it that makes you so absolutely right, and Sir Isaac and me so absolutely wrong?'

Winslow's face flushed. 'Because I believe in God with all my heart. I need no proof. Your Royal Highness, these men are a threat to natural law and justice as written in the Bible. There must be absolute belief or England will slide into chaos. The King's powers are curtailed, and Parliament is still a fragile thing–'

Caroline silenced him with a raised hand. 'My father-in-law, the King, is a devout believer, Mr Winslow, as am I.

There will be no decay in England during our reign.' Her words restored some calm. 'Now, I believe Mr Newton was going to tell us about how he has proved the existence of God.'

Halley stepped back from Winslow and noticed that Newton looked utterly calm. He seemed to have unknotted himself and his head was up. When he spoke it was with assurance, as if the interruption had given him the chance to collect his thoughts.

'Thomas Aquinas believed that the Creation was a continual process; that God's love sustains the Universe around us. I can prove that this is true, yet I am accused of attacking religion. My revelation was universal gravity, a mathematical description of how every celestial object exerts an attractive force on every other body, but that concept leads to a grave paradox. All the stars are surely other Suns, blazing as brightly as our own but viewed from greater distance. They sit in the unchanging tapestry of the night, as proved by the ancient civilisations seeing the same patterns and handing down myths. Yet those stars must all be attracting each other with gravity. With nothing but a simplistic application of my notions, the Universe should have collapsed long before now, everything pulled into everything else.'

Caroline cocked her head.

'Yet the Universe remains stable around us. This, then, is observational evidence of God's sustaining love. I have found the agency through which God nurtures us. He supports the Universe, allowing it to live. Gravity and its actions are God's will, nothing less than the physical manifestation of His given grace. For too long humans have wallowed in doubt, fear and superstition. The revealed scriptures served us to start with, but now we need more. Ever since Martin Luther's Reformation, Europe has been sinking deeper into a crisis of what and how to believe. We need certainty to guide us in this uncertain age. I have shown a method of investigation and proof through a crucial experiment that leads to certain knowledge. Natural philosophy can protect us from going astray. It can give us new gospels to read and believe, new gospels to know the Creator through. Judge me on this and this alone.'

'Double-talk!' spat Winslow. 'You have perjured yourself with this nonsense, if not committed outright blasphemy.'

'Mr Winslow, you grossly overstep your mark.' Caroline's voice was sharp.

'With respect,' snapped Winslow, though his voice held none, 'you have only the makings of a stable monarchy. Until you are established beyond doubt, we need stability of the law as well. Parliament is not a religious body, and it will wrest power from the Church as it has wrested it from the Crown. And these men' – Winslow swung an accusing arm – 'with their science of proof will become a weapon against natural belief. How can you, a woman, possibly comprehend this?'

'I comprehend considerably more than you think, Mr Winslow. There will be no arrests today unless you persist in your disobedience.'

Halley noticed a shiver of fear cross the spymaster's face.

'If any doubt remains of my pure intentions,' said Newton, 'ask others, Your Royal Highness. The noted clergyman Samuel Clarke recently translated *Opticks* into Latin. He is ably positioned to provide an independent assessment of these matters. Ask his advice before you allow others to judge me. Ask him to write to Mr Leibniz on my behalf. I fear that the enmity between myself and your former teacher would not be conducive to impartial debate.'

Caroline nodded. 'Very well, but if I am to leave this room satisfied, Sir Isaac, there is one thing that you must promise me you will do.'

'Anything.'

'Cease your attacks on Mr Leibniz and write him a letter of apology.'

Newton bowed low to disguise his irritation. 'Immediately, Your Royal Highness.'

'Come, Mr Winslow. We've detained these men quite long enough. They are free to go.'

Winslow looked incredulously at Caroline for a moment. Halley noticed how old and tired he looked.

Only when the royal party's footsteps had faded did Halley release the breath he had been holding. Newton flung himself into an adjacent chair, looking more defeated than triumphant.

'Why do people find it so hard to understand what we do?' Halley asked.

'There will always be suspicion where there is ignorance. Ordinary people do not comprehend that when a natural philosopher says *I believe* it is a statement backed by experiment and observation, not a whim or a flight of fancy. It's a wholly new route to knowledge. I hope that even ordinary people will learn this, but it will take time.'

'Then let us pray for that day to come. Perhaps when the comet returns in 1758, and people see the power of your mathematics and method they will be persuaded.' Halley smiled hopefully.

Newton raised an eyebrow at the suggestion. 'Indeed. However, confidence in us and our methods will not be helped if word gets out that we were dragged to the Tower under suspicion by the royal family.'

'Then let us never speak of this scene again,' suggested Halley.

'Absolutely,' said Newton. 'It can remain a fiction, and we will let the Reverend Clarke's letters be our defence in these matters. I'll see to it that he begins the correspondence, even if Her Royal Highness does not request it.'

A moment of understanding passed between them.

Halley clapped his hands to his thighs and stood. 'I must get home to Mary, she'll be worried; but before I go, tell me one thing: does it trouble you that as individuals we'll never know it all? How Nature works, I mean. When I calculated the return of the comet, I knew that it was so far in the future I should never know if I were right or wrong.'

For perhaps the first time, Halley saw an emotion akin to empathy in Newton's eyes.

'To explain all of Nature is too difficult a task for any one man, or even for any one age. Our lives have seen a beginning of something, not an ending. It's much better to do a little with certainty and leave the rest for those who will surely come after us.'

Halley nodded in acknowledgement. 'So, what you said about God's will supporting the Universe – have we really found the boundary of our science, or do you think someone will come after us and explain it with rational means?'

Newton's eyes flashed. 'Only time will tell.'

44

The mice were scratching in the rafters again. Bent double from the stocks, Fatio had kindled a fire in the grate and was now hobbling from one untidy corner to another, collecting together the various pages of his manuscript.

He looked at the sheaf of papers, his unfinished explanation of gravity. His gaze moved to his damaged hands and a lump appeared in his throat. When his wrists had been trapped in the wood, he had wondered if he would ever recover from the numbness. When the pain had flooded back into them after his release, he had wished for the dead feeling to return again. Like his hands, the rest of him had also erupted in agony. He could not stand, sit or lie without pain. Now the only thing he could imagine easing his body was death itself.

Newton's words rang in his head. They were as excruciating as the sensations in Fatio's limbs. *Burn it. Burn the manuscript.*

He squeezed his eyes shut and willed himself to drop the work to its destruction. Yet if he dropped it whole, it would snuff the little flames. Instead, he would have to feed it into the fire piece by piece.

As he held the first pages near to the flames he bit his lip. It was true that he never wanted to see the work again – it evoked too many bitter memories that were waking inside him even now, eating through him like acid – but to destroy the work . . .

At the foot of his bed was a trunk, its metal padlock more secure than the one on his apartment door. He opened the lid and pushed the contents aside to create a gap. Then he laid his manuscript to rest on the pale wood of the trunk's base. After piling other books over it, he turned the key in the lock.

Who – Fatio wondered – would see the manuscript next?

Epilogue

The question of whether Isaac Newton's work was irreligious was forcefully debated in a series of letters between his appointed champion, the English philosopher and cleric Samuel Clarke, and Gottfried Leibniz through Caroline, Princess of Wales. The correspondence ceased with the death of Leibniz in 1716 but was seen as so important that Clarke published it almost immediately. It remains in print to this day as *The Leibniz–Clarke Correspondence*.

Newton never wrote the requested letter of apology to Leibniz, who died in disgrace, branded a plagiarist. Today, however, the German philosopher is recognised as an independent co-inventor of calculus; indeed, it is the notation of Leibniz, rather than of Newton, that is used by mathematicians.

In death as in life, Robert Hooke was overshadowed by Newton. Recently, however, there has been a renewed interest in celebrating Hooke more openly for his wide-ranging contributions to science and architecture.

Newton died in 1727. Towards the end of his life he began to tell the story of how watching an apple drop from a tree inspired him to work on gravitation while at Woolsthorpe during the plague year of 1665. Lauded for his services to his country, Newton was buried in a magnificent tomb in Westminster Abbey. People soon began visiting his birthplace at Woolsthorpe, Lincolnshire, now a National Trust property open to the public.

Edmond Halley's prediction that the comet of 1682 would return in 1758 came true. Spotted on Christmas Day by a

German farmer, the comet's reappearance proved to be a hugely successful 'crucial experiment'. It ushered in the widespread acceptance of Newton's work and methods, transforming them into central pillars of the Age of Enlightenment. The comet is now named after Halley and was seen again in 1835, 1910, and 1986. It will next return to our skies in 2061.

John Flamsteed collected all three hundred unsold copies of Halley's *Historia Coelestis Britannica* (Britain's Celestial History), cut out all the pages he wanted to re-use, then burned the remainder in Greenwich Park. He called the book-burning a sacrifice to heavenly truth. However, he died in 1719, not living long enough to complete his own version. His widow published it for him posthumously in 1725. When Margaret Flamsteed learned of Halley's appointment as her husband's successor, she stripped the Royal Greenwich Observatory of all its telescopes.

Halley was sixty-three when he became the second Astronomer Royal, but lived long enough to record the position of the Moon for an entire nineteen-year Saros cycle. He outlived his wife Mary by six years. On 14 January 1742, aged eighty-five, he drank a glass of red wine and passed away quietly. In his will he asked to be buried next to Mary at Saint Margaret's Church in Lee, London.

Following his ordeal in the stocks, Nicolas Fatio de Duillier left England to wander Europe as a pilgrim. He eventually returned and claimed his manuscript on the cause of gravity was lost, but it was found in his papers after his death in Worcester in 1753. It was purchased by the Swiss physicist Georges-Louis Le Sage, who popularised the concept. However, he spread the idea under his own name rather than Fatio's, and to this day it is referred to as Le Sage gravity.

It piqued interest well into the nineteenth century, when one of its champions was the influential Victorian physicist Lord Kelvin (William Thomson). By this time the re-investigation of gravity had assumed great importance because better telescopes were showing that the planet Mercury did not follow the route prescribed by Newton's laws. Similarly, the question of why the Universe was not collapsing because of its combined gravity was also nagging.

Although no one could make the mathematics of Fatio's idea work, the resurgence of interest in the nature of gravity captured the imagination of many, including the young German-born physicist Albert Einstein.

Einstein's story will be told in *The Day Without Yesterday*, the final book in the trilogy *The Sky's Dark Labyrinth*.

Acknowledgements

Restoration England was a fascinating place, not just because of the seismic shift towards science that was taking place. Religion and politics were in a state of explosive flux, too. My task in trying to bring it all to life would have been impossible without the work of all the authors, historians and scientists who preceded me. Their published non-fiction accounts of these people and events would make ideal reading to move on to if you are curious about where I found my inspiration.

I heartily recommend: *The Leibniz–Clarke Correspondence* by H.G. Alexander (ed.), *The Calculus Wars* by Jason Bardi, *John Flamsteed* by John L. Birks, *England's Leonardo* by Allan Chapman, *Isaac Newton* by Gale E. Christianson, *Newton's Tyranny* by David Clark and Stephen P.H. Clark, *Edmond Halley* by Alan Cook, *The Coffeehouse* by Markman Ellis, *Restoration England* by Peter Furtado, *Halley in 90 Minutes* by John and Mary Gribbin, *The Fall of Man and the Foundations of Science* by Peter Harrison, *The Curious Life of Robert Hooke* by Lisa Jardine, *Newton and the Counterfeiter* by Thomas Levenson, *Newton's Notebook* by Joel Levy, *Genius in Eclipse* by Colin A. Ronan, *A Brief History of Gresham College 1597–1997* by Richard Chartres and David Vermont, *Never At Rest* by Richard Westfall and *Isaac Newton: The Last Sorcerer* by Michael White.

This was a watershed moment between science and religion because it was the first serious suggestion that the new investigation of nature was somehow anti-religious. Even during the trial of Galileo, as readers will know, there was no direct suggestion that the act of studying the Universe was opposed to religion. It was more Galileo's interpretation and his approach to disseminating the information that were judged.

The challenge I faced was how to dramatise an exchange of letters. As I am not Helene Hanff, I decided that the best way was to use my fictitious character, Winslow, and create a scene – the kind of scene that could plausibly have led to the subsequent exchange of letters.

So, please do not think that Halley and Newton were in fact dragged to the Tower and cross-examined. It is, as Newton says to Halley afterwards, 'a fiction'. Nevertheless, the points they debate in that scene are those that Clarke and Leibniz chewed over in the letters. They are the very points that helped to drive a wedge between science and religion.

In portraying my characters I decided not to shy away from their sexuality. This was the Restoration, after all, with England released from the tyranny of the Puritans. The precise nature of Isaac Newton's association with his roommate Wickens, and later with Fatio, is unknown. However, letters exist between Newton and Fatio that are couched in the most intimate and affectionate of terms. While this suggests attraction, there is no evidence to prove a sexual relationship. I have therefore chosen to portray Newton as a repressed homosexual.

I have also taken a conservative approach to Hooke and Grace. There is no doubt that Hooke and his niece shared a sexual relationship – Hooke recorded the events in his diary – the debating point is whether this extended to intercourse. For this story, I have assumed not.

As for the scandal surrounding Halley and Mrs Hevelius, no one knows the truth. All we know for certain is that he did purchase dress fabric for her upon his return to England, and that years later, at Oxford, he was dogged by the rumour of his cuckolding Hevelius. I've given him the benefit of the doubt.

Finally, grateful thanks go to everyone who has helped me by believing in or working on this project: Hugh Andrew, Neville Moir, Alison Rae, Jan Rutherford, Kenny Redpath, James Hutcheson, Sarah Morrison, Vikki Reilly, Anna Renz, Caroline Oakley, Peter Tallack, Duran Kim, Hamish Macaskill, Anna Rantanen, Kim McArthur, Devon Pool, Anne Ledden, Kendra Martin, Ruth Seeley, Nic Cheetam, Stephen Curry.

Also, an enormous thank-you to the thousands of people who bought the first book. To those who came to see me at signings and lectures in bookshops and literary festivals around the world, I am extraordinarily grateful.

Always last, never least, to Nikki.